To Ralph,
With affectio[n]
for forty five years of friendship .

18 Nov. 1983   Alex. Todd .

A time to remember

A time to remember

THE AUTOBIOGRAPHY OF A CHEMIST

—

ALEXANDER TODD

CAMBRIDGE UNIVERSITY PRESS

Cambridge

London New York New Rochelle

Melbourne Sydney

Published by the Press Syndicate of the University of Cambridge
The Pitt Building, Trumpington Street, Cambridge CB2 1RP
32 East 57th Street, New York, NY 10022, USA
296 Beaconsfield Parade, Middle Park, Melbourne 3206, Australia

© Cambridge University Press 1983

First published 1983

Printed in Great Britain by the University Press, Cambridge

Library of Congress catalogue card number: 83-5172

British Library cataloguing in publication data
Todd, Alexander
A time to remember.
1. Todd, Alexander   2. Chemists – Scotland – Biography
I. Title
540'.92'4   QD222.T/

ISBN 0 521 25593 7

Contents

Preface

In recent years it has fallen to my lot to prepare biographical memoirs of a number of eminent Fellows of the Royal Society and in doing so I have been struck with the paucity of information obtainable about their lives and careers, although at least some of them had been associated with major scientific discoveries and had risen from quite humble beginnings to positions of power and influence. It is comparatively easy to track down speeches and lectures given by individuals, and their contributions to science are recorded in the technical literature. Based on such material one can discuss the impact of a person's work and views, but can give little impression of the man himself or the events which influenced his career. In the light of my experience in seeking such information about others, I decided that I should place on record an account of my own life and career in the hope that the pathway from childhood in Glasgow to a Nobel Prize, the House of Lords, and the Presidency of the Royal Society might prove of some interest. Hence this autobiography.

I would emphasise that I have eschewed detailed discussion and argument about my participation in, and views about, public affairs during the latter part of my career; these are matters more appropriate to a different type of book. I have, of course, included, where appropriate, comments on a variety of problems and institutions with which I have been concerned, and I have added as appendices extracts from my Anniversary Addresses to the Royal Society and my Presidential Address to the British Association for the Advancement of Science. I am indebted to these bodies for permission to reproduce the Addresses which set out at some length my views on a number of issues of public concern. In the main, however, what I have written is a general account of my life until about the end of 1980, with emphasis on events which have shaped my career. I have described these events, including their lighter as well as their more serious aspects, as accurately as I can remember them. Memory is, however, a fickle thing and with the passage of time details can become blurred, or even confused. I do not think there are serious inaccuracies in my narrative, but if I have misquoted anyone I hope he or she will forgive me.

My thanks are due to many colleagues and friends who have encouraged me to write these memoirs, and especially to my wife whose patience and memory are both much better than mine, and to Miss Susan Brownell for the skill and effort which she devoted uncomplainingly to the preparation of the manuscript for publication.

Alexander Todd
Cambridge
October 1982

1

Early days – school and university in Glasgow

—

I am often asked when I first became interested in science and when I decided to become a chemist. Questions like these are almost impossible to answer, partly because early memories are patchy and highly selective, and partly because one cannot put precise dates to such things.

I was born on the second day of October 1907 in a rather superior red sandstone tenement block known as Newlands Crescent in Cathcart, a southern suburban area of Glasgow. My father, Alexander Todd, was at the time of my birth a clerk in the head office of the Glasgow Subway Railway Company; in due course he became cashier and secretary of the company, a position which he occupied at the time it was taken over by Glasgow Corporation in 1922. Some time after the takeover, he left to become managing director of the Drapery and Furnishing Cooperative Society Limited, whose main base of operations was a substantial department store at Glasgow Cross. He had long been an enthusiastic supporter of the cooperative movement but was strongly opposed to its political affiliation with the Labour Party or, indeed, with politics in any form. His society, known locally as the D. & F. Stores, which paid a ten per cent discount to all comers, was permanently at loggerheads with the politically oriented cooperative movement in Scotland. My father's family originated in southern Scotland and was settled in the area around Strathaven. My knowledge of it is rather sparse because my paternal great-great-grandfather at about twelve years of age was, with his older brother, deserted by his parents at Glasgow Cross in the early nineteenth century and left to fend for himself. This he evidently did with some success for, two generations later, my grandfather was in business as a jobbing tailor living in modest circumstances on the fringe of the Gorbals area where my father was born. My maternal grandmother (née Ramsay) was the daughter of a farm worker on the Duke of Hamilton's estate at Cadzow in Lanarkshire and came to seek work in Glasgow. There she met and married Robert Lowrie, a foreman in an engineering works at Polmadie where they set up house. In Polmadie my mother (Jane Lowrie) was born within a mile or so of my father's birthplace.

Both my parents were ambitious and hard-working people. My father had only an elementary education and went to work in a Glasgow office at the age of thirteen; from then onwards he was effectively self-taught, apart from some attendance at night classes (none of them concerned with science). By sheer hard work he climbed steadily upwards from this very modest beginning. In all this he was aided and supported by my mother who, likewise, had no more than an elementary education; I believe she worked in a shoe-factory in Glasgow before her marriage to my father. She was a remarkable woman, devoted to her family and backing up her husband in his career. Their story could doubtless be paralleled by many in Scotland; they were determined to battle their way upwards out of the grim surroundings of their youth and succeeded in moving into what might be called the lower middle class. They had a passionate belief in the value of education and were determined that their children should have it at whatever cost. Our family consisted of my sister Jean (died in 1924), five years my senior, myself and my younger brother Robert, born in 1912.

The Glasgow Subway – one of the oldest of underground railway systems – was operated in my childhood by cable traction, and one of my most abiding memories is the characteristic tarry smell of the cable which pervaded not only the stations but also the company's office where my father worked and which was located above the St Enoch Square station. To this day the odour of creosote fills me with nostalgia.

My earliest memory is, however, of a rather scantily clad black woman stirring what appeared to be a large pot of porridge suspended over an open fire; this must have come from a visit to the Glasgow Exhibition of 1910 where one of the more elaborate exhibits was an African village. I have very few other clear recollections of early childhood apart from my entry to Holmlea Public School in Cathcart in 1912 when I was placed in the kindergarten department, only to be removed to a higher form after a few days – more, I fear, because of my physical size than my mental precocity. Our home was situated within a quarter mile of the school but this convenient juxtaposition did not last for long. By early 1914 the family's circumstances had (through my father's efforts as a freelance insurance agent and house factor alongside his Subway job) improved, and we bought a new house about two and a half miles south of Cathcart in the village of Clarkston to which it was planned to extend the Glasgow Corporation tramway system. By the time war broke out in August 1914 and stopped further extension, the trams had been extended from Cathcart to Netherlee, but there they stopped, leaving about a mile and a half of open country road lined with hawthorn and dogrose hedges before one reached Clarkston and our home. Throughout the war years I traversed this road daily on foot in each direction, proceeding onwards by tram from Netherlee to my school in Cathcart, and each day took sandwiches for my lunch in a collapsible metal box. I still remember that road – the misery of walking in boots of very poor war-time quality in the depth of winter, suffering from extremely painful chilblains (doubtless a tribute to the appalling war-time diet of those days), but perhaps even more vividly those dilatory homeward journeys in summer along a dusty road with the hedges aflame with wild roses and convolvulus. All this ceased when, in 1918, I sat for and gained entrance to Allan Glen's School in Glasgow, to which I then travelled daily by train from Clarkston. At Allan Glen's I entered Form Q – the junior school – where I passed the Scottish Education Department's Qualifying Examination and in 1919 passed into the senior school. In this context I might mention that many years later, in the 1950s,

I derived great amusement from the agitation in England about the iniquity of the 'eleven-plus examination' which was supposed to be having such adverse effects on English children. It seemed to me odd that this examination – admittedly under a different name – should have been in operation for about half a century in Scotland without apparently causing any trouble; but perhaps the Scots were a tougher breed.

In those days, Allan Glen's School occupied an old red sandstone building and an adjoining temporary wooden structure in North Hanover Street, about a hundred and fifty yards north of George Square in the centre of the city. The school had no dining hall and we therefore had to eat at lunch time in small restaurants or cafés in the area. I seem to remember being allowed a shilling a day with which I was able to buy a cheap lunch (less lavish than my parents no doubt intended) and leave enough over for hot chestnuts or ice-cream which, according to the season, were obtainable from the Italian street vendors who peddled these delicacies outside the school. One of my earliest recollections of this period is the announcement of the armistice on 11 November 1918. This was the occasion for celebrations in the city, but the school was not given a holiday to mark the occasion. As a result the pupils staged a one-day strike. Militancy is thus hardly a new phenomenon in schools! I cannot remember in detail the reaction of the school authorities – they were probably wise enough to act with restraint.

Allan Glen's School carried as its subsidiary title Glasgow High School of Science. It had been founded in 1853 under the will of Allan Glen, a Glasgow carpenter who believed in the importance of science, and considered that it could be the vehicle of a liberal education just as effectively as the arts. To mark their breakaway from the classical tradition, the first governors decreed that no Greek was to be taught in the school, an instruction which was still being obeyed at the time when I was a pupil. For the rest we were taught all the usual subjects, although mathematics, physics and chemistry were increasingly emphasised as one proceeded through the senior school. I think I was sent to Allan Glen's in preference to one of the other similar schools in Glasgow because I had some vague idea that I would like to take up medicine and was already showing some interest in science. The interest in science, and especially in chemistry, developed rapidly after I went to Allan Glen's, but all enthusiasm for a career in medicine vanished in the spring of 1919. At that time, while tree-climbing with some friends, I fell from an upper branch and dislocated my left elbow rather badly. The joint was set by our local doctor but unfortunately it locked and I found I was unable to straighten my arm. Various rough and ready measures were tried to straighten it without very much success and finally our doctor, with the aid of a colleague who held me down, proceeded to apply main force. As no anaesthetic was used, the operation, although partly successful, was very painful. I decided that if being a doctor meant doing such things to people I would have none of it!

I cannot recall when my interest in chemistry began but it was certainly some time before I went to Allan Glen's. I remember being given a 'Home Chemistry Set' when I was eight or nine years old and I suppose this might have started me off. It was in a pink cardboard box and contained little pill-boxes of sulphur, iron filings, charcoal, etc. with which one could make ferrous sulphide and various kinds of incendiary materials. Once at Allan Glen's my interest grew apace and with it my experimental ventures, the latter reaching a good way ahead of my detailed knowledge, which was, of course, increasing in my school classes at the usual slow and stately pace. The main school building was inadequate to house all classes, and my form used to do practical chemistry in an outlying annexe about half a mile away in Renfrew Street. This annexe was located almost directly opposite the Glasgow premises of Baird and Tatlock Limited, the laboratory furnishers. I quickly discovered that they were quite prepared to sell to me not only chemical glassware, Bunsen burners, and so on, but also (which was more surprising) all sorts of exciting chemicals from concentrated nitric and sulphuric acids to carbon disulphide and chloroform. On such delights I spent quite a bit of my pocket (and lunch!) money and was able to carry out some, fortunately unsuccessful, attempts to prepare nitroglycerol, and at the same time to ruin the stair-carpets at home by dropping acid on them.

My career at school calls for no special comment. I had no great difficulty with any of the subjects except art. I was perhaps the worst performer in the school when it came to free-hand drawing. So bad was I that the art master on one occasion drew attention to my initials (A.R.T.) and observed that my parents 'certainly had a sense of humour'. We were for the most part well-taught in chemistry, which was made a live and interesting subject by the chief chemistry master, by name Robert Gillespie. I at least found the physics teaching much inferior, and I fear that as a result the subject appeared to me to be rather dull and uninspiring although not particularly difficult. I have always regretted this, for it did, in some measure, affect my attitude towards physical chemistry; it is not so much what is taught as how it is taught that determines one's attitude to a subject. In my experience over many years as a university teacher, I have often been struck by the fact that most of the really able young people I have known have been good at most subjects. I suspect that, in many cases, the choice of, say, chemistry instead of botany or indeed of languages as a specialty rests as much on the quality of earlier teaching as on natural inclination.

In the spring of 1924 I passed the Scottish Higher Leaving Certificate examination in English, French, mathematics, physics and chemistry, with German and dynamics as additional subjects at intermediate level. The school liked to encourage boys going to university to remain for a further year in the sixth form before matriculating, but I decided not to do so but to go directly to the University of Glasgow. It may be that I was unusually mature for my years, but at any rate I must confess that I never regretted this decision – I doubt very much if I would have gained anything by a further year at school. I thought it would be a good thing to save my father some money by getting a scholarship of some sort, so I began by sitting for the University Entrance Scholarship offered by the school. This was meant for those boys who had done the extra sixth-form year and the examination was set accordingly. I found that I could not attempt a single question in the physics paper; needless to say I did not get the scholarship. I then decided to apply for one of the bursaries offered by the University of Glasgow in a competitive examination which I duly sat. It was customary to publish the names of the first hundred candidates in order of merit and to make awards accordingly; in the 1924 competition when the list was published my name appeared in the first twenty. But my labours were in vain; had I taken the trouble to study the entry form before filling it in, I would have discovered that there were no bursaries on offer for which I was eligible. I then discovered that the Carnegie Trust for the Universities of Scotland made awards to native Scots which defrayed a large part of their university fees. I accordingly obtained the appropriate application forms. When I showed them to my father, he looked through them and almost exploded. Pointing to one of them in which the parent of the applicant was asked to sign a statement that without assistance he would be unable to send his son to university, he roundly declared that I should know better than to accept charity, let alone be so stupid as to think that he would do so under any circumstances. He then tore up the forms and threw them on the fire. So that was that! I therefore matriculated a couple of days before my seventeenth birthday at the University of Glasgow as a pensioner to read for the degree of Bachelor of Science with honours in chemistry. It is only fair to add that at the end of my first year I was awarded the Joseph Black Medal and the Roger Muirhead Prize in chemistry which did in fact provide me with a scholarship for the rest of my course.

At the time I entered the university, the school of chemistry was adequate if old-fashioned from a teaching standpoint, and not unduly distinguished in research. That is an opinion based upon hindsight, of course, for, at the time I entered, I had no idea whether the school was good or bad. In this respect I differed little from most other school-leavers then and now; in my experience, in non-academic families the choice of university, where it is not made simply on proximity grounds, is usually determined by schoolmasters or, where there is an academic family tradition, by parents who tend to select the institution they themselves attended. It is indeed difficult to see how it could well be otherwise. In 1924, the first-year course for honours chemistry in Glasgow consisted of two terms of lectures on general and inorganic chemistry by G. G. Henderson, the Regius Professor of Chemistry, followed by one term of organic chemistry given by T. S. Patterson, who was Professor of Organic Chemistry. Both professors were good lecturers and the courses given were interesting. This is more than could be said of the accompanying practical class which was devoted to quantitative (mainly gravimetric) and qualitative inorganic analysis. We were given virtually no preliminary instruction apart from one lecture demonstration, and then set to determine the amount of silver in a given solution gravimetrically. If the result obtained was more than two per cent in error, the demonstrator in charge simply wrote 'Repeat' in large letters across one's laboratory notebook and handed out another solution. When, eventually, one got a satisfactory answer, one proceeded to repeat the operation successively with lead, copper, arsenic, bismuth and so on through the traditional sequence of elements and acid radicals used in qualitative analysis. True, we had a few lectures on analytical chemistry on the side, but their relevance to the course was obscure. This probably taught us to become proficient in quantitative analytical procedures, but it was a soul-destroying business. My first – and only – personal encounter with Professor Henderson during my first year was on a November afternoon in 1924 when he was making a tour of the laboratory. He paused at my work bench, looked at the name-label on it, and said 'Ah! Mr Todd, and what are we doing today?' I told him I was endeavouring to determine the amount of silver in a given solution having already failed in three previous attempts. He shook his head sadly, said 'Too bad, too bad!' and passed on his way. I well remember debating with myself whether to abandon chemistry after my fifth 'Repeat' on silver! However, I weathered the storm and went on.

Our second year was given over to organic chemistry which

I found fascinating, both as to lectures and laboratory, and after a third rather dull year given over to physical and inorganic chemistry – with (believe it or not) a further bout of quantitative analytical chemistry as applied to materials containing more exotic elements and radicals – we were able to devote our fourth and final year to one of the three branches of the subject, the practical course including a small research topic on which a thesis had to be written. I chose organic chemistry and, perhaps because I had shown up prominently in all the examinations in that branch, I was put under the professor's supervision. T. S. Patterson's field of research was optical activity. This he pursued in somewhat desultory fashion, since his primary interest was in the history of alchemy, a subject which had occupied most of the attention of his predecessor Ferguson (known for some obscure reason as 'Soda') who had bequeathed a large collection of works on the subject to the university library. It was through Patterson's insistence that the history of chemistry was made a compulsory subject for all undergraduates in their final year, and I have always been grateful to him for thus introducing me to it.

The topic allotted to me for my final year research project was the action of phosphorus pentachloride on ethyl tartrate and its diacetyl derivative. The object was to see whether the nature of the group to be replaced had any influence on the course of the Walden inversion. Needless to say, such results as I got threw no light on that problem, although they did lead to my first publication in the *Journal of the Chemical Society*. Patterson was interested in optical rotatory dispersion and sought to interest me in it also. I read up most of the available literature and began some work on the rotatory dispersion of mannitol and its derivatives which we subsequently published. In June 1928 I graduated B.Sc. with first-class honours in chemistry, being placed first in my year, and was awarded a Carnegie Research Scholarship of £100 per annum to continue research with T. S. Patterson.

I thus returned to Patterson's private laboratory in the autumn, and continued work along the same lines as before. This I did with some diffidence, for I was already getting uneasy about optical rotatory dispersion as a subject for research. For one thing, I did not find it very exciting; I really wanted to do natural product work, holding as I did to the Berzelius definition of organic chemistry (the chemistry of substances found in living matter) rather than the alternative one due to Gmelin (the chemistry of the carbon compounds). More serious, however, was the fact that the subject, as then pursued, seemed to me to have no theoretical basis and was unlikely to acquire one without the application to it of a great deal more mathematical insight than I, T. S. Patterson or even the Lowry group in Cambridge (the other British workers in the field) possessed. It was difficult to see how a junior research student like myself could break loose from it and remain in the Glasgow chemistry department. Furthermore, apart from T. S. Stevens, there appeared to me to be no member of the Glasgow staff with both enthusiasm for organic chemistry and real research ability with whom I would really have liked to work. The others – or so it appeared to me – got on with teaching and made only perfunctory bows in the direction of research. To cut a long story short, by the end of the autumn term of 1928 I had already decided that, if I wanted to make my way in organic chemistry I must leave Glasgow and go elsewhere. Somewhat to my surprise, when I spoke to T. S. Patterson about my feelings he agreed fully, and offered to help in every way he could. I accepted his view that it would be valuable to spend a year or two abroad, if only to learn how other people lived and to acquire real command of a foreign language. Not surprisingly, in view of his own background, he wanted me to go to Paris; I on the other hand wanted to go to Germany, where there was more going on in the natural product field, and in due course I prevailed. The problem, of course, was to decide where to go. In those days, Windaus in Göttingen and Wieland in Munich were the big names. Their laboratories, however, were crowded with foreigners (especially Americans) and English tended to be the *lingua franca* among the research groups. I was anxious to get as much German as I could as well as chemical experience, and my choice finally lighted on Walther Borsche at the University of Frankfurt a.M. Borsche, a pupil of Wallach, had been an associate of Windaus and did natural product work; the other organic professor in Frankfurt was Julius von Braun, well known as a reaction chemist. The set-up seemed to suit me well, so I wrote to Borsche and was accepted to start with him as a research student in October 1929. While all this was going on, my restlessness and decision to leave Glasgow began to spread to other young research students like an infection, and three others decided to do likewise – A. L. Morrison (later Director of Research, Roche Products Limited), T. F. Macrae (later Director of Research, Glaxo Laboratories Limited) and A. Lawson (later Professor of Chemistry, Royal Free Hospital Medical School, University of London). Of these (who were all one year senior to me in Glasgow) Morrison went with me to Frankfurt, while Macrae and Lawson went to Munich at the same time; in passing, I should add that we none of us ever regretted the move, and we have remained lifelong friends as a result of it.

I find it difficult to give an objective view of the University of Glasgow as it was in my undergraduate days. One's first experience of university life always becomes a treasured memory, and like most features of one's youth the unpleasant aspects tend to be forgotten and the whole experience seems to exist in a kind of rosy glow. One remembers, vaguely, student activities – the Union, the Charities Rag, the preposterous elections to the office of Lord Rector and the rowdy installation of that dignitary. I do not recall being overworked; I played a lot of tennis, although mainly for one of the city clubs rather than the university, and I acquired a good acquaintance with the dance halls, theatres and football grounds of the rumbustious and at times violent city of Glasgow. As to the actual academic courses, I have already said something about chemistry, in which they were almost wholly factual and where we heard practically nothing about the new electronic theories of organic chemical reactions, although these were already part of the regular courses as near to us as St Andrews! But the actual teaching was good, except in physical chemistry, and I owe a great debt of gratitude to the late T. S. Patterson for all the help and encouragement he gave me. In the undergraduate course we had to take a number of subsidiary subjects; I took physics and mathematics in my first year, geology in my second, and metallurgical chemistry in my third. In my third year I also took, as a voluntary extra subject, bacteriology. Of these subsidiaries I have little of consequence to say. Looking back now I recall particularly the efforts made by the Professor of Mathematics to get me to read his subject for honours; this was flattering, but was based on the misapprehension that, because I hadn't much difficulty with the course, I must be interested in the subject! Geology I found fascinating – so much so that I once contemplated taking additional honours in it. It gave me two things of great importance. Firstly, a knowledge of palaeontology which was my introduction to biology and, secondly, an ability to look at landscapes in the light of geology, which has greatly increased my appreciation of travel all over the world. Metallurgical chemistry I found useful as providing me with some knowledge of heavy industry and of the practical application of chemical, and particularly physico-chemical, theories – e.g. the Nernst heat theorem – which had hitherto seemed to me rather abstract. I also learned a great deal more about the phase-rule and its applications through the iron–carbon diagram, which was the central topic of our lectures on steel, together with quite a lot of rather useless factual material like the flow-sheet of the copper smelters at the Union Minière du Haut Katanga. Both metallurgical chemistry and bacteriology I took in the Royal Technical College in George Street, Glasgow, never suspecting for one moment that, some thirty-eight years later, it was to become the University of Strathclyde and that I would be its first chancellor.

When I graduated in 1928 jobs were hard to come by and the majority of my contemporaries went into the teaching profession. For those of us who wished to do research it was also rather a difficult time, for research grants or scholarships were few and not of a very substantial character. I was lucky to be awarded a Carnegie Research Scholarship (£100 p.a.). The Department of Scientific and Industrial Research in those days offered a few awards of £120 p.a. but it was firmly believed in Glasgow that there was a prejudice against Scottish applicants. Whether or not there was any truth in this I do not know, but I never met anyone in Glasgow whose application for one of these awards had been successful.

In my day there was (and I believe there still is) an active society embracing staff and student members of the chemistry department called the Glasgow University Alchemists' Club which organised lectures on chemical topics as well as social activities. During the latter part of my student days in Glasgow I was much involved in the club's operations and, indeed, it was at one of its meetings in session 1928–9 that I read my first paper to the Alchemists' Club – I suppose my first lecture ever – on the topic 'Did Paracelsus wear spats?' The paper was taken with moderately good grace by T. S. Patterson, although I fear it was a rather juvenile 'take-off' of one of his star lectures on the career of that mediaeval rogue. During my undergraduate years the club rather ambitiously ran a magazine entitled *The Alchemist*; as I recall it, the only year in which it actually made a profit was that in which I served as its business manager and shamelessly used my father's position in the city to bully a number of local firms into buying advertising space. The magazine, I am told, continued to exist for many years until rising costs eventually forced its demise in the sixties.

The Alchemists' Club apart, there was relatively little communal life in Glasgow University when I was there, since most students lived at home. It may well have been different for those who lived in the very few residential hostels, but for most undergraduates social activities tended to centre on their home environment; in this respect most non-collegiate universities were probably alike. Nevertheless, one learned a lot from contact made with people of widely varying interests in the university and outside it, and from a thorough acquaintance with the common people of Glasgow – their hopes and aspirations as well as their virtues and vices. From that experience I think the most important thing I learned was that tolerance is one of the great virtues and that hasty judgements are only rarely sound.

2

Apprenticeship in research – Frankfurt and Oxford 1929–34

Although neither of my parents ever had more than an elementary school education they were firmly convinced of the value of education and, in the belief that I had some talent, they saw to it that, even if it meant some sacrifice on their part, I went to a good school and later, when I was ready for it, to university. In this they were strongly encouraged by my uncle Walter F. Todd. The latter was very much younger than my father, being a child of my grandfather's second marriage. Orphaned about the time I was born, he stayed for some years with our family while he studied at Glasgow University with a view to taking up teaching as a career. This career he never followed for he enlisted in the army in August 1914 and after serving in Gallipoli and the Middle Eastern campaigns he became a professional solider and retired shortly before the Second World War as a staff colonel in the Cameronians. It was, however, quite a shock to my parents to learn in 1929 that I proposed to study in Germany. Neither I, nor they for that matter, had ever been out of the United Kingdom and they viewed foreigners with great suspicion. However, they accepted my decision without protest, although my mother had grave doubts as to the level of civilisation I would find in Germany; so much so, indeed, that she plucked up her courage and insisted that my father should take her to Frankfurt a few weeks after I got there, just to satisfy herself (which she did) that Germany was a tolerable place for her son to live in!

Before I went to Frankfurt in the autumn of 1929 the

Carnegie Trust increased the value of my scholarship to £150 p.a. which made my plan to study there much easier to carry through. This sum, it is true, was barely adequate but with the Reichsmark at twenty to the pound sterling one could manage tolerably well on about £200 p.a. I found plain but comfortable lodging in Königstrasse hard by the university in Bockenheim at, I think, 35 Rm per week including breakfast; other meals could be got at the 'Schlagbaum', a noisy hostelry at the Bockenheimer Warte, for as little as 50 pfennig. Having moved into my room, I reported to Professor Walther Borsche at the Chemical Institute of the university. My knowledge of the German language was at this stage rudimentary and Borsche spoke very little English. Our conversation was, accordingly, much restricted but we established that I would take up a problem in the bile acid field – the nature of apocholic acid – and that I could have a certificate from him, to transmit to the university administration, that I knew sufficient German to understand the lecture courses. I need hardly say that the certificate grossly exaggerated my linguistic capability but, armed with it, I duly matriculated and registered myself as a candidate for the Doctor phil. nat. in organic chemistry with physical chemistry and mineralogy as subsidiary subjects. The need for a certificate of proficiency in German was, incidentally, the direct cause of my first meeting with Bertie Blount who did his undergraduate work at Oxford and arrived just a few days later than I did, to work for his doctorate with Borsche on the constituents of kawa root. On the forenoon of my second day in the laboratory I was sorting out my glassware (everything except retort-stands and the like had to be purchased by students) when Borsche (a small grey-haired man invariably attired in a white laboratory overall which was too tight for his ample figure) came in and with some difficulty explained that he had an Englishman in his room and desperately needed an interpreter – would I come and play this rôle. I then went with him to his office where, reclining in an armchair, was Bertie. I asked him what the trouble was, whereupon he said, 'There really isn't a problem. I have been trying to tell the old boy I need a certificate of proficiency in German but I can't get him to understand.' When I had translated, Borsche roared with laughter – and promptly wrote out the required certificate. This meeting with Bertie Blount began what has been a lifelong friendship.

The Frankfurt laboratories were quite an eye-opener to anyone coming from Glasgow or, for that matter, Oxford. Organic micro-analysis was being done as a routine service, catalytic hydrogenation using Skita-type colloidal platinum and palladium catalysts at room temperature and atmospheric pressure was normal practice, while glass apparatus with standard interchangeable ground joints was in widespread use; these, with many other gadgets like Jena sintered glass filters and so on, were unknown in the laboratory I had come from and, I suspect, in virtually all British laboratories. Borsche was a good person to work with – a good experimentalist and a patient supervisor. He had little enthusiasm for theory being a typical example of the classical German organic chemist. He was, however, completely devoid of the arrogance shown by many of his contemporaries, and indeed it has always seemed to me that it was his very gentleness and his patent desire to avoid strife that prevented him from earning a more prominent place in German science than he in fact occupied. I had some personal experience of this in my doctoral work. Among the various dehydration products of cholic acid, apocholic acid, the structure of which I sought to elucidate, was something of an anomaly. My work on it led me in due course to recognise that it was impossible to ascribe to it any structure compatible with the then accepted carbon skeleton of the sterols and bile acids. That skeleton had been proposed by Wieland and Windaus who had received the Nobel Prize for chemistry in 1928 in recognition of their massive work on bile acids and sterols respectively, and both of them were held in the greatest respect by Borsche, who had himself been an associate, and remained a tremendous admirer, of Windaus. When, towards the end of 1930, I told Borsche that, in my opinion, the Wieland–Windaus structure for the bile acids must be wrong and suggested an alternative which would accommodate my results with apocholic acid he was much disturbed. He pointed out that this would imply that there were errors of fact or of interpretation in the work of Windaus and that it would be presumptuous of me to suggest such a thing. It was only after much persuasion, aided perhaps by news that the accepted formula was suspect in Munich also, that he finally agreed to publish my work in *Zeitschrift für Physiologische Chemie* (1931, 198, 173). The structure which I proposed embodied a seven-membered ring and was, of course, erroneous, although, oddly enough, in its favour I pointed out that the skeleton could be derived from farnesol and two hexose molecules! I was probably too young and inexperienced to press on to the proper answer on my own, and Borsche felt – perhaps wisely, and certainly generously – that we should let Windaus have our results and leave any further follow-up to him. So it was that, when I completed my doctorate and returned to England, I left the sterol/bile acid field and never returned to it.

Julius von Braun, the Director of the Chemical Institute, was a very different type from Borsche. Also rather a small man, he had a rather round bullet-cropped head, atop a very short neck. He had a rather florid complexion which matched his aggressive manner; rather incongruously he had a rather high-pitched squeaky voice. He was known variously in the laboratory as '*der edle Pole*' and '*der Bonze*' and his assistants as well as his doctoral students (who described him as '*katzen-freundlich*'), were certainly made to toe the line. He could be, and usually was, quite single-minded in his approach. In my time in Frankfurt he had a group of students working on the structure of naphthenic acids which occur in petroleum. I remember in particular one of these students called Gradstein, who seemed to spend most of his waking hours ozonising naphthenic acids down in the departmental cellar where von Braun kept a large ozone generator. One afternoon there was a substantial explosion in the cellar – loud enough to cause considerable alarm even up in my laboratory on the second floor. We promptly rushed downstairs, and pushed open the cellar door; there was a certain amount of smoke about, and the unfortunate Gradstein was lying flat on the floor apparently unconscious (it transpired later that he had merely fainted).

I and the others were about to go to his aid when von Braun arrived, pushed everyone aside, stepped over the prostrate Gradstein and said '*Ja, und die Substanz?*' There must be many stories about von Braun. I did not have a great deal to do with him myself, nor did my colleagues Blount and Morrison. For one thing, I don't think he really approved of our working with Borsche rather than with him and, for another, I rather blotted my copybook with him quite early on before my German was fluent enough to permit of diplomatic or evasive answers. One day at the beginning of the *Wintersemester* 1929 he asked me if I would come along to the seminar he ran for his research students. I asked when it was held and he told me it was on Saturday at 8 a.m. I suppose I should really be ashamed of myself, but I am afraid I told him in my very blunt German that I was busy on Saturdays, and that in any case 8 a.m. was far too early for me. Perhaps it is not surprising that our relations thereafter were rather cool. Nevertheless, whatever his faults, von Braun was a brilliant reaction chemist and I learned quite a lot from him during my stay in Frankfurt. With von Braun and Borsche on the organic side, Schwarz in inorganic, Dieterle in pharmaceutical, and, across the road, Bonhoeffer in physical chemistry we had a strong and on the whole harmonious chemical school. It is sad to think that all of them fell foul of, and suffered under, the Nazi regime in later years – not, as far as I know, because of Jewish ancestry, but because they had independent minds.

I had completed enough experimental work for my doctoral thesis by Easter 1931 and I returned then to Scotland to shake off the after-effects of a severe attack of influenza. I returned to Frankfurt a couple of months later to submit my thesis and take my Dr. phil.nat. examination. The examination was oral and was in three parts: (*a*) a one-hour oral on organic chemistry, (*b*) half an hour on physical chemistry and (*c*) half an hour on mineralogy, the examination in each section being conducted by the appropriate professor in the presence of an umpire – in my case the professor of physics who occupied the time by reading the *Frankfurter Zeitung*. Although the examination was public, one never had an audience of more than two or three and these were simply men who were intending to take their examination shortly and wanted to see what kind of punishment to expect. There was a curious ritual associated with the *Doktorexamen* in Frankfurt (and probably elsewhere in Germany) at that time. About a week before the date of the examination one called on each of the examining professors at his home bearing a bunch of flowers for his wife, and had a chat. Although the chat was largely given over to pleasantries with the professor of your main subject – in my case Borsche – the visits to the other professors were much more important. It was accepted that a candidate would probably not be expert over the whole area covered by his subsidiary subjects and so one was expected to indicate those parts in which one was most interested (i.e. knew something) and those in which one was less interested (i.e. knew nothing). This pretty custom was intended to, and usually did, avoid embarrassment to both parties in the examination. As a matter of fact it didn't work out too well for me in physical chemistry. Bonhoeffer had come to Frankfurt – his first full professorship – at about the time that I did; being in any case a rather absent-minded young man, and still rather unaccustomed to examining, he confused what I had described to him a week before as interesting and uninteresting. Accordingly, I had a rough half-hour on pre-dissociation spectra and some other photochemical topics based on a lecture course through which I had usually slept since it was given twice weekly at 5 p.m. in the summer semester after I had been swimming in the *Stadtwald* since lunchtime. However not too much damage was done since, at the end of the day, my 'Note' for the thesis was 'Ausgezeichnet' and for my oral examination 'Sehr gut'. My other subsidiary, mineralogy, had been chosen by me partly because I had some little acquaintance with the subject from my Glasgow course in geology, and partly because it shared with bacteriology the reputation of being the easiest option in the faculty; bacteriology lectures, however, were given in the hospital located on the other side of the city which made my final choice a simple one.

In opting for mineralogy I chose better than I knew. It turned out that Nacken, the professor in Frankfurt, was a great admirer of Gregory, the geologist in Glasgow, and seemed to consider any pupil of Gregory as a positive asset to his department. Accordingly Morrison and I were welcomed with open arms and our pathway through the subject was made very easy; Bertie Blount also benefited from this for, although he knew no mineralogy at all, it was assumed that, since he came with me to be interviewed, he too must be a Gregory pupil!

I was much impressed by the oral examination for the doctorate which, especially in the principal subject, was quite searching. The technique – which was customary and did not apply simply to my case – was to ask a question of the candidate and see whether from his opening reply he really knew something about the topic at issue. If he did, the examiner passed rapidly to another topic; if not, the candidate was subjected to more probing so that his knowledge or lack of it could be ascertained. In this way it was possible to cover a very great amount of ground in the course of an hour and an experienced examiner could find out much more than he could have done by means of written papers. One is often told how students in German universities tended to wander from one university to another during their courses. This may well have been true in some places, but it did not seem to be the case in Frankfurt save in the following special circumstances. It was generally held that, if it became clear to you that you were likely to fail in your examination, the proper thing to do was to transfer to the University of Giessen for your last semester; this was in fact done from time to time by the weaker vessels.

Looking back now, my recollection is that my life as a student in Frankfurt was a very happy one. No doubt matters were helped by the fact that, having apparently a certain facility for languages, I acquired fluency in German rapidly and so was able to get to know people much better than would otherwise have been possible. The student body was pretty cosmopolitan comprising, as it did, many impoverished young men from Eastern European countries belonging to the old

Austro-Hungarian Empire, with a sprinkling of Russians, both *emigrés* and Soviet engineering students, Persians and other Middle Easterners with a few Americans but, as far as I could ascertain, no British students other than ourselves. Frankfurt had taken over the colours and traditions of the German University of Strassburg including its student corporations, although it still retained its designation '*Preussiche Staats-Universität zu Frankfurt a.M.*' So it was that, although duelling was illegal, Frankfurt, in my day, still had several of the old fighting corporations. My friend Morrison had rooms in the same house as one Otto Löchel, president of one of these, the '*Freie Landsmannschaft Franco-Saxonia*', and we became quite friendly with him. In due course we were admitted as honorary '*Freie Landsmänner*' and used to attend from time to time the weekly *Kneipe* or club meeting complete with our song books and ribbons all in the orange and silver colours of the *Landsmannschaft*. These meetings were devoted mainly to singing and beer drinking and were relatively peaceful, although I recall on one occasion there was a row over some girl which led to a sabre duel. Duelling occurred at rather infrequent intervals and was usually the traditional '*Mensur*' in which the contestants' sole object was to give one another a few facial scars. I did attend one of these events. It was held at 5 a.m. in the yard of an inn in Sachsenhausen, round which we had a series of scouts posted on the look-out for police. This particular '*Mensur*' had been arranged with the Darmstadt branch of the Franco-Saxonia which had sent up two or three of their latest recruits (*Füchse*), all of them freshmen at the *Technische Hochschule*, to swap scars with our recruits. It was a curious performance – almost a ritual – each duel being stopped by the attendant doctor as soon as a suitable face wound had been inflicted. At the end of it all we repaired to the inn and despite the early hour consumed vast quantities of beer before returning to Frankfurt. Nevertheless, contacts such as these with '*Korpstudenten*' from the various corporations put me much more in touch with the student body than would otherwise have been possible. Most of the students I knew were right wing in their political stance although few of them were extreme. For the most part they sympathised with the democratic National Party; it was only later as the Weimar Republic finally began to collapse that they moved over mainly to the National Socialist Party. Franco-Saxonia apart, we lived the normal student life. In those days Frankfurt was an attractive city with its spacious west end and its dignified central area within the *Anlagen*: south towards the river lay the beautiful *Altstadt* with its *Weinstuben* which we used to frequent when we were feeling moderately affluent. Alas, all that has gone since the Second World War, and the Frankfurt I knew has been replaced by a rather featureless modern city with hardly any of its old character.

The period from 1929 to 1931 was an interesting time to be in Germany. It began with the winding up of the Allied occupation – I remember celebrating the departure of the French from Mainz with my student friends – and ended with a financial crisis involving, in Frankfurt at least, a run on the banks. During these two years, even I was aware that the Republic was crumbling, and there was widespread and growing economic distress. It seemed that government neither could nor would do anything and, hand in hand with growing disillusion and cynicism about parliamentary democracy, the extreme parties – National Socialists and Communists – began rapidly to make ground. After a good deal of violence on the streets, the Nazis gained the upper hand, the middle ground of politics fell away, and most of the public, including the student body, gradually moved over to them. This was not because they had any real sympathy for the extreme anti-Semitism of the Nazis; anti-Semitism in a mild form was, and I imagine always had been, widespread in Frankfurt but, alas, none of the men I knew really took the Nazi fulminations seriously. What the happenings during those years taught me was that economic troubles coupled with weak and vacillating government leave the way to totalitarianism wide open; all that is needed to complete the disaster is the appearance of a brilliant demagogue such as Adolf Hitler undoubtedly was. And it could happen in any country.

When I went to Germany I had no clear ideas about a career

except that I wanted to do research in the natural product field, but soon it was time to think about what to do after completing the doctorate. Clearly I wanted to go back to Britain sometime but not necessarily at once. Bertie Blount was of similar mind and we did make an abortive attempt to go to Moscow to work for a spell with Zelinsky; it was probably well for both of us that it failed, for by all accounts life was pretty hard for students in the Soviet Union at that time. Blount had also been pressing me to go to Oxford with him and, since in 1930 Robert Robinson had succeeded W. H. Perkin Jr there, I decided to do so provided I could find the necessary financial support. During my short period in Glasgow in the spring of 1931 (when I did a little work on hesperidin) at the suggestion of Professor T. S. Patterson I applied for an 1851 Exhibition Senior Studentship. Entries for this had to go through one's own university, which then nominated its chosen candidates to go forward for the main competition. I remember well a somewhat discouraging interview with the Registrar of Glasgow University who said he would forward my application but that nothing would come of it since Glasgow hadn't figured in the award list for many years, and, moreover, there was an applicant in another subject who was much senior to me and would be the university's first choice. However, somewhat to my surprise, I was in fact awarded a Senior Studentship and I thereupon made arrangements to join Oriel College and start research with Robert Robinson in September 1931 on the synthesis of anthocyanins (the red and blue colouring matters of flowers). Although by that time I already held the German Dr. phil. nat., it eased the problem of entering a college if one read for a degree; since the D.Phil. course at Oxford had only a research requirement and was unlikely to influence adversely any work I intended to do, I enrolled accordingly and, indeed, took the degree in 1933.

When I went to Oxford people in my position, i.e. graduates from other universities coming to do advanced work, often found great difficulty in fitting fully into college life, and not a few of those who were my contemporaries never really became part of the colleges which they joined. In Oriel I was much more fortunate, became thoroughly integrated and, indeed, came to be regarded as an 'Oriel man' just as much as those who had come up as undergraduates. This made a great difference to my social life in Oxford and, since, after taking the D.Phil. in 1933, I was made a member of the Oriel Senior Common Room I really saw a lot of college life both of the student and the don. It is perhaps only fair to say that my ready integration into Oriel really rested on sporting ability. When a newcomer came to the college the various clubs used to descend upon him to see if he had any qualities that would be useful to them. Now, it happened that I had played quite a lot of lawn tennis since my schooldays, and, although no champion, I was competent enough to have played tournament and representative tennis. I confessed to the Oriel captain that I had played a little tennis and was promptly asked to go out to the sports ground where the college experts would give me a trial. As it happened, I thrashed the college experts and was immediately accepted as a worthwhile member, a position made secure by my election to the University Lawn Tennis Club in my first term. But whatever the reason, I had three very happy years in Oriel.

Robert Robinson had come to Oxford just a year before I joined him, but already the Dyson Perrins Laboratory was a hive of activity. He had brought with him a group of research men of various nationalities, some pre- and some postdoctoral, and there was an air of excitement in the laboratories. Robinson himself was at the height of his powers and bubbled over with ideas. I had never met him until I went to join him in September 1931 and I still recall with amusement our first encounter in his office. He was seated at his desk writing, and when I came in he looked up and said 'So you've come to do research?' whereupon I replied simply 'Yes.' 'Well,' he said, 'you know I am interested in anthocyanins and for our synthetic studies we need various ω-hydroxyacetophenone derivatives; now I would like you to make some veratroyl chloride and see if you can convert it with diazomethane to a diazoketone and thence to the hydroxyketone.' He then picked up his pen, evidently finished with me. As I was still standing there he looked up and, seeing my rather puzzled expression, said 'What's the trouble?' 'That doesn't sound much of a problem,' I said, 'what about the research I have to do?' He spluttered a bit. 'What? Who are you?' 'My name is Todd.' 'Good God,' he said, 'I thought you were a Part II undergraduate!' He then started to chat about work and took me off to instal me in a small laboratory for two people adjacent to his own, where, incidentally, he also installed my Frankfurt friend Bertie Blount who was to work in the alkaloid field.

Blount and I occupied that laboratory for the next three years and, probably because of its proximity to his office and laboratory, we saw a lot of Robinson – much more than did most of his collaborators elsewhere in the building. We used to brew tea at about four in the afternoon and on most days Robinson would drop in and join us for a concerted attack on the *Times* crossword puzzle and a gossip about current chemical interests. It was during one of these 'tea and puzzle' sessions that I made the real breakthrough that opened the way to the synthesis of all the major diglucosidic anthocyanins. When I got to Oxford, one of the main stumbling blocks in efforts to synthesise such flower pigments as, for example, cyanin, pelargonin and malvin, was the total failure that had attended all efforts to make the 2-glucoside of phloroglucinaldehyde. Robinson was thinking up some very roundabout procedures to get out of this difficulty, but it seemed to me that since one could put a benzoyl group directly into the 2-position of phloroglucinaldehyde one ought to be able to put in a glucose residue. So I started in to condense acetobromoglucose with unprotected phloroglucinaldehyde; I tried all sorts of tricks, but I could get nothing but intractable syrups and gums. One day, however, I had a methanolic solution of one such gum in a small conical flask and was concentrating it by dipping it from time to time into a hot water-bath while at the same time having tea and doing the crossword puzzle with Robinson and Blount. The inevitable happened; I dipped the flask in for a little too long so that the solution boiled violently and, probably because it was rather hot, the flask slipped from my fingers and fell into the water-bath. I fished it out and put it over among the dirty glassware which I intended to clean up the following morning; to my astonishment when I came to do this the next day I noticed that the walls of my small flask were covered with crystals. They were indeed crystals of 2-beta-tetra-acetyl-D-glucopyranosyl-phloroglucinaldehyde! Henceforth, with seeding material available, there were no further problems and the way to the anthocyanins was wide open. I don't know what moral, if any, to draw from that story, but at least it explains why, in a later edition of a well-known chemical textbook, it is recorded that the 2-glucoside of phloroglucinaldehyde is best crystallised by diluting its concentrated methanolic solution with a large volume of hot water!

Compared to the Frankfurt Chemical Institute the Dyson Perrins Laboratory in Oxford was rather primitive when I joined it in 1931. No micro-analysis facilities were available, and analyses were carried out down in the semi-basement by the laboratory steward, Fred Hall, using the classical macro-procedures. Fred had been Perkin's chief assistant and in my time ruled the Dyson Perrins – and its professor – with a rod of iron; many stories are told about him but I always got on well with him and found him very helpful. [I did hear, years later, that he told someone that the only two real gentlemen who ever worked in the Dyson Perrins were Alex Todd and Donald Somerville (who later went into the law and politics and became Attorney General).] Blount and I used to send our materials to Schoeller in Berlin for micro-analysis and it was not until 1933 when Drs Weiler and Strauss came as refugees from Nazi Germany that routine micro-analysis was developed in the Dyson Perrins. Blount and I also introduced catalytic hydrogenation and ground glass joints to Oxford but, during my time, the other equipment available amounted to a very indifferent polarimeter, a visible absorption spectrometer and an American pressure hydrogenator which didn't work. Robinson was like his teacher and predecessor Perkin in having little interest in gadgets – he was firmly attached to the degradative and synthetic methods of classical organic chemistry and was slow to adopt such things as ultra-violet and, later, infra-red spectroscopy as aids in structural work.

In those days Robinson was at the height of his powers and he was the most inspiring director of research with whom I have ever come in contact. Certainly one had to be reasonably tough and independent to appreciate him fully, and many a budding chemist who came to Oxford from another school where 'spoon-feeding' of Ph.D. students was the rule, found it difficult to settle into the rather haphazard Oxford scheme of things. Robinson had a razor-sharp mind, but he was interested in many topics and his interest would flit from one to the other with great frequency. He was liable to concentrate all his attention on the topic interesting him at any given moment, to the exclusion of everything else. This made many people regard him as frequently tactiturn if not downright rude, and his collaborators had to get accustomed to being alternately badgered about their progress several times a day, and being almost totally ignored for weeks on end. Perhaps because of the proximity of my laboratory to his, I only observed but did not suffer much from his behaviour. Robinson did not pursue solid experimental work, and differed in this way from his predecessor Perkin. Generally he confined himself to a few preliminary experiments, usually in glass boiling-tubes, and left the follow-up to a junior collaborator. He was very emotional in his reaction to events and impatient with those holding views contrary to his own. This perhaps helps to explain the enormous range of his contributions and the reason for his name being associated with the discovery of a prodigious number of reactions used in synthesis, but with relatively few completed syntheses of complex molecules like steroids. His instinct when confronted with a difficulty in experimental work was at once to seek an alternative route to his objective, or even to change the objective itself; this practice led frequently to the discovery of new reactions, but also, at times, to the premature abandonment of synthetic routes which were later shown by others to be practical.

My sojourn in Oxford was a very important period in my career. Quite apart from establishing a permanent bond with Robert Robinson I learned to know and made lifelong friends among the host of young chemists who, like me, came to the

Dyson Perrins to work with him, or followed him there from elsewhere – Gulland, R. D. Haworth, Baker, King, Erdtman (Sweden), Sugasawa (Japan), Walker, Schlitter (Switzerland), Morf (Switzerland), Ramage, J. D. Rose, Briggs (New Zealand), Watt (Australia) and many others. We also had two refugee professors from Germany during the latter part of my stay – Arnold Weissberger and Fritz Arndt, whose chain-smoking of cigars ensured that his presence in the laboratories was always well advertised!

In 1932 I completed the synthesis of the flower colouring matters hirsutin, pelargonin, malvin, and cyanin, chlorides, and so effectively rounded off the anthocyanin field leaving, as far as synthesis was concerned, only some mopping up operations. Accordingly I was looking around for some new subject when Harold Raistrick asked Robinson if he would like to look at the chemistry of some red pigments in certain plant pathogenic moulds of the *Helminthosporium* group; the problem was turned over to me and much of my time was devoted to it during my last two years in Oxford. Before getting immersed in that problem I had prepared a quantity of gossypol from cottonseed, but never got around to studying it; indeed, I rather think I still have the material in a bottle in my specimen collection in Cambridge. I dropped it in favour of a very brief foray into the steroids following the first (erroneous) structure advanced by Rosenheim and King in 1932. I began to check certain degradations of cholesterol but abandoned them when the correct structure was put forward shortly afterwards by Wieland and Dane. Thereafter, it is true, I devoted a few weeks to a rather hare-brained scheme whereby I sought to generate a tetracyclic nucleus of the steroid type by a very short route beginning with a somewhat unlikely reaction between hexatriene and methyl cyclohexenone. The first problem was to prepare hexatriene; the literature route *via* divinylethylene glycol was rather unattractive. While I was doing some preliminary work on possible modifications, Robinson travelled with Professor J. F. Thorpe to Manchester on a consulting visit to the Dyestuffs Group of Imperial Chemical Industries and mentioned my problem to him. Thorpe immediately said there was no need to synthesise hexatriene since it was readily available as a major constituent of the so-called 'railway hydrocarbon', the low-boiling residue left behind in the gas cylinders still used for lighting railway carriages in those days. All that you needed to do was to warm it slightly and pass the gas evolved into bromine whereupon you would obtain a copious supply of hexatriene hexabromide. It was characteristic of Robinson that, before returning to Oxford, he ordered a quantity of railway hydrocarbon from the London Midland and Scottish Railway Company. The stuff was duly delivered in a thirty-gallon metal drum which was dumped in the laboratory yard where, it being high summer, it lay gently hissing and smelling to high heaven. It was closed by a large hexagonal nut which appeared to be immovable. However, aided by the laboratory handyman, a monkey wrench and a hammer, I forced an entry. With difficulty we capped the resulting gusher and managed to collect some of the contents. Following Thorpe's instructions I then used up every particle of bromine in Oxford but obtained only a mixture of ethylene dibromide and butadiene tetrabromide contaminated by – at most – a trace of the hexatriene compound. I returned to the preparation of divinylethylene glycol and the railway hydrocarbon – at the urgent request of the local inhabitants – was returned to the railway company.

3

Edinburgh, London and Pasadena

My 1851 Senior Studentship was extended for a third year and was due to expire on 30 September 1934. By the early part of that year I had therefore to begin casting around for something with which to support myself. It was not an easy time for finding an academic position, which I would have preferred to one in industry, although I did not rule out an industrial career; but industrial jobs were about as scarce as academic ones. I remember being approached about taking charge of a laboratory at the Courtauld Institute of Art which was of no interest to me, and also about an assistant lectureship at Bangor which I did not find at all attractive. However, in May 1934 George Barger, then Professor of Medical Chemistry in the University of Edinburgh, visited Oxford to seek Robinson's advice. Barger, who himself was half Dutch, was a friend of B. C. P. Jansen of Amsterdam who had in 1926 first isolated the anti-beriberi vitamin B_1 from rice hulls in Batavia. Jansen, who did not consider himself a chemist, had heard that both the IG-Farbenindustrie in Germany and Merck & Co. in the United States had embarked on the structural investigation of vitamin B_1, with a view to its commercial synthesis, and he felt that he really ought to do something himself. He therefore wrote to Barger and asked if he would take up the problem. Barger was naturally interested but felt he was not really qualified for the job since it would involve working with very small amounts of material – all that was available from Jansen was about five milligrams of crystalline vitamin and a description of the isolation procedure from rice (which contains a few milligrams per ton). So he came to Oxford to ask Robinson what he should do. Robinson suggested he should ask me since I was a natural product chemist and I had acquired micro-techniques in Germany. I jumped at the offer, since it gave me the chance of doing exactly the kind of work I wanted to do; vitamins were just becoming accessible to organic chemists and I was fascinated by the possibility of finding out why they were so important, i.e. what function they performed in living creatures. So I went to Edinburgh in the summer of 1934 on the basis of a Medical Research Council grant eked out by the promise of some part-time demonstrating in medical chemistry; this meant, overall, a cut of around twenty-five per cent in my income but I reckoned it would be worth it.

Barger was a close friend of Marcus Guggenheim, the Research Director of Hoffman La Roche & Co. of Basle, and through him was in close contact with that firm. As a result the firm agreed to do the preliminary concentration of vitamin B_1 for me and send to Edinburgh bottles of the concentrate which, corresponding in weight to less than 0.1 per cent of the original rice hulls, made it possible for me to isolate small amounts of vitamin for study; without that help the scale of operations needed would have been impossible for an academic laboratory. Perhaps this would have been even more true in Edinburgh than in some other places, for the Medical Chemistry Department was not only inconveniently housed in the massive Victorian pile of the medical school but, when I arrived, the general level of equipment was deplorable. To aid the vitamin B_1 work Barger got a substantial grant from the Rockefeller Foundation and with it I was able at least to ensure that we got adequate glassware and some minor instruments. But I was not able to spend more because of Barger's extraordinary attitude to the grant; for some obscure reason of his own, he felt it was up to him to show how economical he could be, and he was determined to return as much as possible to the Rockefeller Foundation. In this he was successful, and, indeed, when I finished the work on vitamin B_1 he spoke with pride of how it had been done with so little Rockefeller money!

I admit that when I first saw the Medical Chemistry Department my heart sank and I was exceedingly depressed. But there was one bright spot which helped to restore my optimism. This was the presence of Dr Franz Bergel. An Austrian by birth, he had been a young *Privat-Dozent* in Freiburg when, following Hitler's accession to power, he came to Edinburgh as a refugee supported by a grant from Hoffmann La Roche obtained through the efforts of Guggenheim and Barger. We quickly struck up a friendship which has lasted to this day, and he decided to join me in the vitamin B_1 work. We did in fact work together for four years, first in Edinburgh and then in the Lister Institute, before we parted in 1938 – I to Manchester University and he to the new research department of Roche Products Ltd at Welwyn Garden City. I soon gathered round me a rather motley group of people – Barger's laboratory was much frequented by foreign students, perhaps because of his wide international contacts. At any rate, I soon had working in the B_1 field not only Franz Bergel but Anni Jacob from Frankfurt (she remained with me until 1944), Juan Madinaveitia from Madrid (who later married Anni), Karimullah from Lahore, Keller from Basle and Fraenkel-Conrat, a German refugee who is now in Berkeley, California.

I need not detail here all the hectic and at times frenzied work during these two years in Edinburgh. Details of the work can be found in the literature. We were beaten to final synthesis, although only by a short head, by the Germans and Americans but our synthesis was not only quite different in concept from the others but proved to be sufficiently superior for it to be used by Hoffmann La Roche to take a major share of the world B_1 market. I should add that, since my work was partly supported by the Medical Research Council, I was not allowed to patent anything. This attitude, I fear, cost the country – and probably me also – very large sums of money and this may have been, in part, responsible for a later change of heart by the Research Councils. It incidentally played a role in the creation, much later on, of the National Research Development Corporation. Through advance knowledge of our results, which we naturally supplied to them in gratitude for their assistance, Hoffmann

La Roche were in a position to establish themselves in the field and to take advantage of the opportunity thus presented to them. I have never grudged them their success, and remain ever grateful for the help they gave me on vitamin B_1 and on vitamin E where I had an exactly similar relationship with the firm. Hoffmann La Roche ever since those days have always been ready and willing to help me with grants or materials required for my research and always without commitment on my part. In my experience their behaviour has been a model for industry/university collaboration in general.

One or two recollections of the B_1 work may be worthy of mention here. The first reveals another trait of George Barger. The first breakthrough in the structural work on vitamin B_1 was made by R. R. Williams and the Merck group, who effected what has come to be known as the 'sulphite cleavage' and which yields a thiazole derivative together with a pyrimidine sulphonic acid. The result first appeared (I think in December 1934) as a paragraph in the *New York Times* in such a form that it was not wholly clear just what had been established. I can't remember now exactly how I got hold of this information, but I told Barger of its existence and he said I should draft a telegram to Max Bergmann at the Rockefeller Institute in New York and ask him to ascertain the facts and let us have some more detailed information. I accordingly drafted a telegram and showed it to Barger who almost exploded. 'Todd,' he said, 'you are an unmitigated spendthrift; that telegram would cost us several pounds! Let me redraft it for you.' This he did, reducing it to about a dozen words and told me to despatch it. The following morning we received in return a telegram from Bergmann which read 'Cannot understand your communication. Please explain.' I then sent off my original version and got the required information much to Barger's chagrin! The delay between then and 1936 in establishing the true structure of the vitamin was due to an error made initially by all three groups working on it. The 'sulphite cleavage' was clearly fission of a quaternary ammonium salt. This was so unexpected that we all jumped to the entirely erroneous conclusion that the pyrimidine ring was directly linked to the quaternary nitrogen and that an ethyl group was attached to the pyrimidine ring system. It took nearly eighteen months for the German group to establish that in the pyrimidine sulphonic acid from the cleavage the sulphonic acid residue was attached through a CH_2-group to position 5 of the pyrimidine ring system. It followed that there must indeed be a CH_2-group between the pyrimidine ring system and the quaternary thiazole in the intact vitamin. Bergel and I, meanwhile, went directly for the erroneous formula by total synthesis, a difficult task in which we succeeded; as the product was not identical with the vitamin we could easily deduce the correct structure (which was one we had always regarded as a possible alternative) and we proceeded to synthesise it. The fact that we first had to synthesise the correct pyrimidine intermediate delayed us as compared with the others and left us a month or two behind the Germans and somewhat less behind the Americans; but on the other hand our prior work on the wrong B_1 structure gave us a much more elegant vitamin synthesis. So, in the end, not a great deal was lost and we certainly could congratulate ourselves on our performance against two large and powerfully equipped organisations. The final clearing up was done after I moved to the Lister Institute in London in the autumn of 1936.

Towards the end of my Edinburgh period I also began preliminary work on vitamin E, although I only developed that work seriously after moving to London. However, my stay in Edinburgh had another, and indeed vital, consequence for my career. It was there that I first met a young lady, Alison Dale, who was doing postdoctoral research in the department of pharmacology under A. J. Clark. Pharmacology was next door to medical chemistry and I fear I spent quite a lot of time there. Suffice to say that by the time I left Edinburgh we were engaged to be married and did indeed marry in January 1937 after I had moved to London. That was perhaps the best thing I ever did, for my wife has always been a vital part of my career; to her I am forever grateful. Her father was Sir Henry Dale, the famous physiologist, and through her and her family I also met many people in the biomedical field, and these contacts have undoubtedly affected many of my scientific interests. (As a good Scotsman I can record that my fiancée consented to become formally engaged while we were attending a meeting of the Biochemical Society in Aberdeen. I at once bought her an engagement ring at Woolworths in that city. Hardly were we back in Edinburgh when one of the 'diamonds' fell out of its setting. Ever free with cash, I told her to throw away the ring and then bought her another – in the Edinburgh branch of Woolworths!)

In the early summer of 1936 it was announced that J. M. Gulland, Reader in Biochemistry at the Lister Institute of Preventive Medicine, had been elected to the chair of chemistry at Nottingham and the head of the biochemistry department at the Lister, Robert Robison, began to cast around for a successor. His enquiries of Barger, Dale and Robinson brought up my name and no doubt I had the attraction of being (apart perhaps from Haworth and Hirst) the only chemist in Britain actively operating in the field of vitamins with which the Lister Institute had long been closely identified. After all, it was there that Casimir Funk coined the name 'vitamine' for the anti-beriberi factor, and Harriette Chick was in command of their large nutrition department. Be that as it may, I was asked to join the Institute in place of Gulland, although it was pointed out by Robison that, at 28, I was really too young to have the title of Reader which was accordingly withheld. I have often wondered if the withholding of the Readership was meant to be a smack in the eye for Robert Robinson who had been asked by the Lister for an opinion on me and my promise as a chemist. Many years later the correspondence relating to my appointment fell into my hands, including a letter from Robinson in which he said he had no doubt that my appointment would be good for the Lister, but was doubtful whether the Lister was good enough for me. These remarks were not well received in the Lister Institute, and perhaps they felt it was time Robinson was taken down a peg! Be that as it may, I was quietly and unobtrusively appointed to a Readership a few

months after moving to the Lister. I moved down from
Edinburgh with Franz Bergel, Anni Jacob and T. S. Work as
camp followers and was soon joined, first by Hans Waldmann
from Basle, and later Marguerite Steiger who came from
Reichstein's laboratory in Basle where she had done synthetic
work on cortical hormones; Juan Madinaveitia also came down
to the Lister towards the end of my stay there. I also brought
my rat colony for vitamin E testing and maintained it at the
Lister Institute. Miss Chick and her colleagues didn't really
believe in the existence of vitamin E when I went to the Lister,
but they provided me with the facilities for keeping rats. With
the help of Miss A. M. Copping and a small grant from the
Medical Research Council we kept the colony going, and in fact
did all the biological assays of vitamin E ourselves. It is only
fair to say that Miss Chick was readily converted once we and
others had isolated tocopherols from rice- and wheat-germ oils
and had shown that they produced consistent results in the
prevention of abortion or resorption of the foetus in pregnant
rats.

The Lister Institute was in those days a curious place. It had
a substantial section devoted to bacteriology, which was not
surprising since the Institute had a branch at Elstree which
produced sera and vaccines in bulk; indeed, the sale of these
materials was the main source of income for the Institute in
Chelsea Bridge Road. Its other main activities were nutrition
and biochemistry, and my group was something of an oddity
since I was much more chemical in my approach than Robison
or his predecessor Arthur Harden. Our habit of working late
at night and at weekends and our production of a wide range
of penetrating and at times not very pleasant odours did not
increase our popularity in an institution which was in any
event rather inward-looking and whose staff, I am afraid,
formed something of a mutual admiration society. Nevertheless, during my stay at the Lister Institute we tidied up the
vitamin B_1 studies by making a number of analogues, isolated
beta-tocopherol (one of the vitamins E) from rice-germ oil,
established the main features of its structure and embarked on
the synthesis both of it and of alpha-tocopherol. In addition we

started work on the active principle of *Cannabis indica* (*C. sativa*) and, with Madinaveitia, on the spreading factor (hyaluronidase) present in testicular extracts. On the whole, then, we accomplished quite a lot during our two year disturbance of the Lister Institute's otherwise peaceful existence!

Our work on *Cannabis* at the Lister brought me into an early and, in retrospect, slightly absurd confrontation with the Home Office Drugs Branch. The starting material for our studies was a distilled extract of hashish which had been seized by police in India and had been obtained from them by my colleague Franz Bergel while on a visit to that country some years before and while he was still resident in Germany. The distilled resin was transmitted to Germany *via* the diplomatic bag, and, in due course, brought to Edinburgh through the port of Leith together with a variety of other chemicals in a suitcase carried by Bergel; no questions were asked by the Customs. In starting our work in the Lister we first isolated cannabinol from the resin, and showed that, contrary to general belief, it was pharmacologically inert, the hashish effect residing in the material left after its removal. We submitted a brief paper on these observations to a meeting of the Biochemical Society early in 1938 and this was duly printed in *Chemistry and Industry*, which in those days published short abstracts of papers read at such meetings. Within two or three days of the appearance of our little note I received a letter from the Drugs Branch inviting me to come to the Home Office and speak with one Inspector X at my early convenience. This interest of the authorities in me and my work was unexpected, but I went to the Home Office and was duly shown to the room of Inspector X who was seated at a large desk on which lay a copy of *Chemistry and Industry* and what looked like a large ledger. After exchanging the usual courtesies the Inspector said 'I see you have been doing some work with hashish,' to which I could only reply 'Yes.'

'You realise, of course, that *Cannabis* in all its forms is proscribed.'

'I suppose it is.'

'Well, we can probably straighten things out fairly easily so don't worry. Here in this book I have a record of all the legal holders of *Cannabis* in this country with the amounts of material they hold.'

I glanced quickly at the opened ledger. On the page visible to me there were a dozen or so names of doctors and professors each with small amounts of drug opposite them – usually only a few grams or ounces.

'Now,' said Inspector X, 'presumably you got your hashish from one of these holders and the only irregularity is that he didn't notify me; but we can easily put that right by an appropriate entry. Which of them gave it to you?'

'I'm afraid none of them did.'

'Then who did?'

'The Indian police.'

'Yes, but how did it get into this country?'

'In a suitcase at Leith.'

'You mean that you smuggled hashish?'

'I wouldn't call it smuggling. It was in a properly labelled flask, but the Customs people didn't seem to be very interested in it; it was just one of a number of bottles of chemical specimens in the suitcase.'

At this point there was a brief silence; then 'How much of the stuff have you got?'

I confess I had been waiting, not without trepidation, for this question and at first I tried to parry.

'Well, of course, what I have is a distilled extract of hashish and not the drug as it appears on the Indian market.'

'Never mind about that – just tell me how much.'

I plucked up my courage. 'Two and a half kilograms.'

'Good God!'

The Inspector looked worried and after a few moments he said 'What are we going to do about this?' followed by a long pause and then 'I think we had better make you a licensed holder of *Cannabis*.'

So he wrote in his ledger 'Dr Todd $2\frac{1}{2}$ kilos', and added 'You will of course understand that this material must be kept under lock and key, that all amounts you use in your work must be duly recorded, and that your records will be open to inspection by us at any time. Furthermore, if you publish any papers arising from work with this resin we will expect twenty-five reprints of each paper.'

'Certainly. Where shall I send the reprints?'

'Send them to me at the Bureau of Drugs and Indecent Publications.'

Until my appointment to the staff of the Lister Institute, I had existed entirely on research awards of various types and had given little thought to such things as security of tenure. My outlook on such matters changed somewhat after my engagement to Miss Dale. She gave up her Beit Memorial Fellowship in the summer of 1936 and returned from Edinburgh to her parents' home in Hampstead; I found lodgings in Gordon Square, Bloomsbury, not far from Warren Street Underground Station, nicely poised for Chelsea and Hampstead between which places I divided most of my time until we were married on 30 January 1937. Already during the summer of 1936, after I had accepted appointment to the Lister Institute but before I had taken up my duties, other possibilities began to appear. In July of that year I, with my fiancée and her parents, attended a garden party at the Robinsons' home in Oxford. There we met President Cody of the University of Toronto who was visiting Oxford and was looking for someone to occupy the vacant chair of organic chemistry in Toronto. Encouraged no doubt by Sir Robert Robinson and Sir Henry Dale, Cody asked me if I would accept the chair; I said I would, and he departed for home a few days later. Some time thereafter I had a long letter from Cody explaining that I was rather young and that it might be easier for me to come as Associate Professor and be promoted to Professor in a year or two's time. My fiancée and I discussed this proposition at great length and, of course, consulted her father and Sir Robert. Finally we agreed that this was reasonable and I wrote off saying I would accept the offer. After some delay I had another letter from Cody saying things were very difficult in the chemistry department in Toronto and offering me the post of Assistant Professor! At this I blew up and wrote a fierce letter addressed to President Cody in which I told him exactly what

I thought of him and the University of Toronto! Sir Henry Dale was somewhat alarmed by the letter when I showed it to him, and asked that he be allowed to send it first to his friend Charles Best (of insulin fame) in Toronto who would decide whether to pass it to Cody. I believe Best conveyed the gist of it but not the actual letter. I suppose he was right, but at the time I felt very sore about it. However, my wife and I had just married by then and were settling in to a flat we had found in Wimbledon, so that such matters didn't disturb us too much.

When we were married we went to spend a day or two in the New Forest, but decided to defer the honeymoon proper until winter was over and go in April to Portofino in Italy. So it was that we found ourselves in April 1937 spending a fortnight at the Hotel Splendido (the title was rather flattering) on the hillside overlooking the tiny harbour of Portofino. There were few other visitors but they included an American couple who looked to be ten years or so older than ourselves, and who kept themselves very much to themselves – as, I suppose, we did. I did not recognise either of them, although I quickly deduced that the husband at least must be a chemist. While swimming at nearby Paraggi, I saw him lying on the beach reading *Chemical Abstracts*; this behaviour I found peculiarly repellent in such surroundings and decided that we would not seek to press acquaintance! It was our custom to have breakfast on the hotel terrace and, while doing so one morning a telegram was brought to me. I opened it and read the surprising contents. 'Are you interested blo-organic chemistry Pasadena. Letter follows. Millikan.' I laughed and then read it aloud to my wife. Out of the corner of my eye I noticed that the interest of the American chemist was evidently aroused by what he had overheard, but we neither of us said anything. I mention the American couple because just over a year later at Harvard we met them again and all four had a good laugh at our recollections of Portofino – they were in fact Professor and Mrs Louis Fieser whose textbooks on organic chemistry became almost universally used.

In due course, after our return from Portofino, I received a long letter from Dr Millikan, President of the California

Institute of Technology, setting out his proposition. Briefly put, Mr and Mrs Crellin, a wealthy Californian couple, had made a substantial benefaction to the CalTech for organic chemistry and with it a new building – the Crellin Laboratory – was being erected alongside the existing Gates Laboratory where the young Linus Pauling had recently been installed, and which was largely devoted to physical chemistry. My kind of organic chemistry was not well represented in the United States in those days, and apparently the Rockefeller Foundation informed Millikan that if he could persuade me to come they would provide $1 million to help develop and maintain the Crellin Laboratory as a department of bio-organic chemistry; I believe that this was the first occasion on which the expression 'bio-organic' was officially used (it certainly was novel enough to fox the international telegraph company!). I was not prepared to accept without first seeing the situation in Pasadena at first hand, and I proposed that my wife and I should go there for a month or six weeks in the spring of 1938 on a basis of no commitment on either side: this was accepted by Millikan, and it was agreed that we should visit California in March 1938. My decision to go was not greeted with marked enthusiasm either by Sir Henry Dale or Sir Robert Robinson. Apparently they were quite happy that my wife and I should go to Toronto, because they thought we would only spend a few years there before returning; but Pasadena was another story, for they feared that if I went there I would probably stay! Perhaps it was for this reason that Sir Robert pressed me to submit an application for the chair of chemistry at King's College London at the end of 1937. This is the only job for which I have ever made a formal application – and the University of London's electors to the chair did not even bother to take up my references; which was, perhaps, just as well, for I really had no desire to go to King's College.

So off we went in March 1938 after giving a promise that we would not formally accept any offer made by Pasadena until we had returned to England and could take our final decision away from the glamour of California. We had an exceedingly rough crossing of the Atlantic in the United States

Lines' ship *Washington* and my first sight of America was snow-
and ice-bound New York. We spent a few days up the Hudson
River near Ossining with Dr H. D. Dakin and his wife. Dakin
was a very old friend of the Dales and was affectionately known
in their family circle as Uncle Zyme. He pursued his researches
in biochemistry in a private laboratory built in the grounds of
his home, and was one of the gentlest and wisest scientists I
have ever known. From New York we went by rail (NYC) to
Chicago where we had to change stations and proceed further
on the Union Pacific Railroad's 'Los Angeles Limited'. On the
train to Chicago we met a friendly and cheerful businessman
returning to his home in Wichita, Kansas after what seemed
to have been a successful trip to New York. He tried hard to
get me to sign a petition aimed to unseat President Roosevelt;
when I pointed out that I wasn't an American citizen he told
me that wouldn't matter, since 'any signature is OK if it helps
get rid of that so-and-so!' We had a few hours to wait in
Chicago before the 'Los Angeles Limited' left in the late
evening, and our casual acquaintance on the train insisted on
showing us around in a car and then took us to the beer-cellar
at the Brunswick Hotel where we spent the evening consuming
sausage and beer with waitresses in Bavarian costume, a
Schuhplattler group for entertainment, and German as the
language used by most of the customers. Our train from
Chicago proceeded through Iowa overnight and reached
Council Bluffs and the Missouri River the following morning,
then wound its way across the desolate winter landscape of
Nebraska, through Wyoming and then into Utah and Salt Lake
City on a fine clear day which gave it a strikingly beautiful appearance when seen from afar. Here we were informed that
severe floods in Southern California had severed the railway to
Los Angeles, and we were given the option of going on to San
Francisco or proceeding by rail to Cedar City, Utah, spending
the night there in the train, and going on next day to Los
Angeles by bus. We chose the second alternative and have
never regretted it. Our bus took us *via* Las Vegas (rather less
brazen than it is today) and across the Mojave Desert *via* Baker
to California. I shall never forget my first sight of the lush

orange groves of Southern California as the bus, which had travelled all day through desert, suddenly emerged from the Cajon Pass to reveal San Bernardino with its green orchards lying in the valley below. We arrived in Los Angeles in pouring rain and were taken across the intervening open country to Pasadena where we lodged in the Athenaeum – the Faculty Club of the CalTech. Fortunately the weather cleared overnight, and my recollection is that it remained fine and warm thereafter during virtually the whole of our five week stay.

In those days CalTech, although growing, was a considerably smaller and perhaps more tightly-knit organisation than it is today. We saw a lot of Linus Pauling and his wife Ava Helen; my wife and I became their firm friends and we spent a lot of time with them – picnics here and there and a camping trip in the desert area south of Los Angeles. I discussed chemistry and its prospects in Pasadena a great deal with Linus who would, of course, have been my opposite number had I in fact gone there, and I spent quite a lot of time drawing up equipment lists for the new Crellin Laboratory and planning both staff requirements and courses. I was greatly helped in considering the outlook and in appreciating the (to me) peculiarities of the American academic scene by J. B. Koepfli, an organic chemist who had worked with Perkin and Sir Robert Robinson and was now an associate of the CalTech, working there as an honorary professor; Edwin Buchman who had been with von Braun in Frankfurt when I worked there with Borsche, and later worked on vitamin B_1 with H. T. Clarke in Columbia University was also there. He, like Joe Koepfli, was wealthy and had a purely honorary position at CalTech. As time wore on my wife and I became more and more convinced that we were likely to finish up in CalTech. Finally, towards the end of our visit, we were invited by Dr and Mrs Millikan to accompany them on a trip south to see the unfinished new observatory which was being built on Mount Palomar (the mirror for the telescope to be erected there was at that time being ground and polished in a special building on the CalTech campus). The trip was most interesting despite some unconsciously daredevil driving on mountain tracks by Mrs Millikan and a rather cold night in the temporary quarters on the mountain top. During the trip Dr Millikan made me a formal and very attractive offer which included two trips each year to the midwest and eastern States and one to Europe every second year (for in those days California was rather isolated scientifically). I told him of my promise to Sir Robert and Sir Henry, and undertook to give him a final answer with no bargaining ten days after my return to England. By this time both my wife and I were pretty certain that we would accept and be back in Pasadena in the autumn.

True, there were certain things that worried us slightly; we were well aware of the mounting danger of war in Europe and we were rather distressed to find so many people in California who were not only unconcerned about this possibility, but felt that poor Hitler was being given a raw deal by Britain and France! All they seemed to be concerned about was Japan – perhaps, as it turned out later, quite properly, although their grounds for fear at the time seemed rather irrational and associated with Spengler's '*Untergang des Westens*' and vague fears of the 'yellow peril'. It seemed probable that, if war broke out in Europe, America might remain neutral, and in such a situation I would probably find it impossible to remain there. However, my wife and I felt that, on balance, we would take the chance.

Meanwhile, back in England I. M. Heilbron, Professor of Chemistry in Manchester, was appointed to succeed the late J. F. Thorpe in the chair at Imperial College, thus rendering vacant the most important position for a British organic chemist after Oxford and Imperial College. Clearly there would be a bit of a shuffle, but as I never for a moment expected I would be a candidate for Manchester, especially after my experience with the King's College chair, I wasn't unduly perturbed. However, it is only fair to say that, on the night before we left Pasadena, Joe Koepfli told me that he reckoned that I would join the CalTech unless I were offered Manchester; such an offer, he felt, I could hardly refuse, as Manchester was much more in the chemical swim than CalTech and he and (he claimed) Pauling and Millikan considered it more than an outside possibility.

Early in May Alison and I, having already had a look around for a possible house, left Pasadena and travelled in leisurely fashion by the Santa Fe railroad to the east coast, making a brief side-trip to the Grand Canyon *en route*. Thence we proceeded to Cambridge, Massachusetts and stayed for some days with Oliver and Alice Cope. Oliver, a young surgeon at the Massachusetts General Hospital, had been associated earlier with Sir Henry Dale and in that way knew my wife. We met a variety of the Harvard people there, including the Fiesers, with whom we enjoyed mutual recognition from the Portofino days! Thereafter we returned to New York and sailed for England on the *Queen Mary*, then the pride of the Atlantic and certainly the largest and most iuxurious ship of the day.

(I look back on the various trips made to and from the United States by ship with nostalgia. The 'Queens' in particular were magnificent ships in every way, and the leisurely days spent on the ocean are an abiding memory. I think the last ocean voyage I made was on the *Queen Elizabeth* about 1954. Air travel has spoiled transatlantic trips!)

We left ship at Plymouth and proceeded to our little flat in Wimbledon where, to my considerable surprise, I found awaiting me a telegram from Sir John Stopford, Vice-Chancellor of the University of Manchester, telling me that Heilbron was going to Imperial College and asking me to come to Manchester on the following Thursday to talk with him about the situation. This I did; I had never before been in Manchester, but I was met at London Road Station and taken to the university. I sat in an anteroom for, I suppose, about fifteen minutes before Sir John appeared, introduced himself and took me along to the Senate Room where, to my astonishment, a group of professors was assembled and it dawned on me that this was in fact a formal interview and that there were probably other 'candidates' there that day also. (This I found later was correct; J. W. Cook and R. P. Linstead had been on the mat before me.) I was cross-questioned about this and that for a quarter of an hour or thereabouts, and then taken by the Vice-Chancellor to his room. By this time I was getting slightly alarmed, and I decided that I really would have to be completely frank with him about my position. I told him that it was clear to me that our 'little talk' had been in fact an interview and that, while I wished to make it clear that I was in no sense holding a pistol at his head, the time available was short. I said I had in my hands a firm offer from CalTech in Pasadena and that I had promised to give Dr Millikan my reply by cable ten days after returning to England, i.e. on Monday, three days hence. I also said that, unless I had an equally firm offer from Manchester before Monday, I would accept the Pasadena offer. Sir John smiled and said 'Don't worry – I quite understand. Sit down with the newspaper – I'll be back in five minutes.' Off he went – and he came back as promised in about five minutes. 'Well,' he said, 'will you take the job?' 'Yes,' said I. And that is how, at the age of thirty, I became Sir Samuel Hall Professor of Chemistry and Director of the Chemical Laboratories of the University of Manchester.

4

Manchester and the move to Cambridge

Having found a house in Withington my wife and I moved to Manchester in late September 1938. It was not a very comfortable month with war-clouds looming in Europe; this was the time of Hitler's annexation of Czechoslovakia and Chamberlain's Munich Agreement. I remember taking a pretty dim view of the situation. Indeed, shortly before we finally quit the Lister Institute I remember being with Franz Bergel (who was just starting up as Director of Research at Roche Products Ltd, Welwyn Garden City) in a pub not far from the Institute and we discussed whether we shouldn't perhaps join the army forthwith rather than wait a few weeks for the actual outbreak of war! However, the worst was not to happen for another twelve months when, as it turned out, we were as chemists kept out of the army to undertake other work of national importance.

At the time I went there, the Manchester school of organic chemistry had a high reputation. The traditions of Perkin, Dixon, Lapworth and Robinson had been upheld by my immediate predecessor Heilbron, who left a well-equipped and progressive department behind him. It was a large school – third in the country after Oxford and Imperial College London (Cambridge, although physically large, had got into a very low state under the ageing W. J. Pope). In those days Manchester was (so Heilbron told me) generally known in the academic world as 'the first-class waiting room'. I could hardly believe that I had been given the job, and I viewed it with some trepidation since I knew that I would be the youngest member of the teaching staff of the chemistry department. I needn't have worried on the latter score – I was welcomed by the whole staff, who were without exception extremely helpful.

(Stopford, tackled in the Athenaeum Club about my appointment, said it was like managing professional football – he would always sign a good man when he saw one, and offered bets that Oxford or Cambridge would be around very soon to try to arrange a transfer!)

Having now got a settled position with a large school and extensive laboratories, I could begin to spread my wings a bit and especially to start something really big even if it might take a number of years and a lot of co-workers to produce any results. My position in research at the time I went to Manchester was roughly as follows:

(1) I had to clear up some of the vitamin E work.
(2) Having found cannabinol inactive and having a supply of *Cannabis* resin I wanted to get on with my pursuit of the active principle.
(3) In continuation of my vitamin interests I was following up a liver filtrate factor of the B complex and also pursuing antianaemic factors in liver extracts.
(4) I had some minor pieces of work, e.g. on *Erythrophleum* alkaloids of which we had a sample obtained many years earlier from Merck of Darmstadt and from which Blount (in Oxford) had isolated the crystalline erythrophleic acid and done some preliminary work on its structure.
(5) Already in my work with vitamin B_1 I had become very interested in the reasons for the importance of vitamins and how they could function. By this time evidence had been accumulating which showed their participation in coenzymes like co-carboxylase, cozymase and flavin-adenine-dinucleotide.

I wanted to go into this field which, certainly from a synthetic point of view, was largely untouched but too difficult and chancy for a young worker without the facilities and scope of a big school like Manchester behind him. Several of the coenzymes including adenosine triphosphate (ATP) contained nucleosides or, perhaps more properly, nucleotides, and there was a lead through them to nucleic acids which, even at that time, I believed were in some way tied up with transmission of hereditary characteristics (from their presence in chromosomes). Here then was just what I wanted – a big field, relatively unexplored, and which, though difficult, could yield in the end results of great importance. I therefore laid out a three-pronged attack: (*a*) nucleoside synthesis; (*b*) phosphorylation (about which astonishingly little was known) and (*c*) nucleotide synthesis covering phosphorylation of nucleosides, coenzymes and possibly later nucleic acid structure and synthesis.

Altogether a tall order, but the field in view had few if any occupants. P. A. Levene had, of course, done quite a lot of work following on the lines of Emil Fischer's early studies, and Gulland was working with a small group on nucleic acids but not as a serious competitor in the directions I wished to pursue. There was thus one great advantage – the literature was small and there was therefore little to read up! (I always tell youngsters looking for a field of research that they should (1) choose an important one; (2) choose one large enough to give room for changes in direction; and (3) avoid fashionable fields and choose if possible one in which they themselves would be the authors of all or most of the relevant literature.)

Such then was my general philosophy, and I put all these lines into operation beginning in the major new field with some preliminary skirmishes in the nucleoside and phosphorylation areas.

I was extremely fortunate to have as my colleague in the chair of physical chemistry that remarkable scholar and scientist Michael Polanyi. Prior to my interview for the Manchester post I had not met him, and on that occasion I had no more than a passing word with him. As soon as my appointment was officially announced I therefore went up to Manchester to meet him and discuss plans for the future. It was a fine summer day when my train drew into London Road Station and there was Polanyi to greet me. We went outside, bundled into his small Ford car, and set off. By the time we reached the foot of the sloping approach to the station Mishi (the name by which I always knew him) was already well embarked on a theoretical discussion which was brought to an abrupt halt when, emerging into the traffic of London Road, we collided with a stationary tramcar. We drove the slightly dented car without further incident to the university, had a brief chat, and proceeded to the Staff House for lunch at the end of which Mishi discovered that he had forgotten to bring any money with him that day; it is only fair to say that the indulgent way in which this confession was received by the cashier and her polite refusal to let me help suggested that this was not the first time such a thing had happened. I tell the story only because it serves to underline the essential unworldliness of Michael Polanyi. Usually deep in his own thoughts (which were always well worth listening to), he paid little attention to everyday life around him. He began life as a medical man, served as such in the Austro-Hungarian army in the First World War and thereafter, in his long career, was, in succession, experimental and theoretical physical chemist, economist, social scientist and ultimately philosopher. He was a truly remarkable man and one whose close friendship I was privileged to enjoy during those Manchester years, when he was moving gradually away from physical chemistry to economics, a move which accelerated after his main Manchester disciple, M. G. Evans, moved to the chair of physical chemistry in Leeds during my stay. Mishi left the running of the chemical laboratories almost entirely to me, which may have been just as well since his administrative technique usually involved issuing on impulse some sweeping directive to Dr A at lunch time and then spending most of the afternoon seeking subterfuges which he hoped might ensure that the directive (now regarded by him as bad) would in some way become inoperative without his having directly to countermand it.

I had the good fortune to be able to make the move from London to Manchester with scarcely a break in my research; I took with me two of my Lister Institute colleagues – Dr Anni Jacob and Dr Marguerite Steiger. Both of them were not only outstanding research chemists but also women of strong character. They quickly organised not merely the young research students but also the technical staff of the laboratory and got them going in their work; no slackness was tolerated! But our quick start and rapid progress thereafter also depended very largely on a new appointment which I made. When Heilbron left Manchester he took with him to Imperial College his laboratory steward F. G. Consterdine. Fred Consterdine was a very efficient person, and was generally regarded in chemical circles as the country's outstanding laboratory manager; he had built up a smooth-running organisation in Manchester, but it was clear that unless someone equally talented and efficient could be found to replace him, that organisation would be unlikely to survive for long. Such a man I found in A. R. Gilson. At the time of my appointment he was a junior member of the Manchester chemical laboratory staff. His potential had not passed unnoticed by Polanyi (nor, as I soon learned, by Consterdine), who mentioned Ralph Gilson to me as a promising youngster whom I ought to look at. I did, and was enormously impressed both personally and professionally. I therefore appointed him laboratory steward, gave him *carte blanche* with my full confidence, and I never did a better day's work. Appropriately enough, he was by far the youngest applicant being about five years junior to me – I being the youngest member of the teaching staff of my department. Ralph Gilson and I struck up not only a partnership but an enduring friendship; between us I like to think that we put, in turn, Manchester and Cambridge on the scientific map and he, over the years, has probably done more than any other man I know to guide the development of chemical laboratories both in design and equipment. Much later our ways parted – but only superficially – when he left university work in Cambridge in 1956 to become managing director of Perkin Elmer (Great Britain) Ltd. I must add that the entire chemistry staff, teaching, research, and technical, gave us their full support right from the start; I shall always be grateful to them and to the Vice-Chancellor, Sir John Stopford, who backed me at every turn and from whose wise advice on administrative problems I greatly benefited. The only occasion when my youth got me into a bit of trouble was when, late one evening a few days after taking up office, I went down to the chemistry department to do some writing. In this endeavour I was unsuccessful, for the night watchman on duty refused me admission on the grounds that students were not allowed in the laboratories after normal working hours unless they had written permission from the Professor!

Manchester University was a lively place in those days. Although I had only one year to see it in peacetime conditions, life in it had zest and vigour even during the war years. It certainly had no fear of appointing young professors. Three other very young men in addition to myself were appointed to chairs in 1938 – Willis Jackson (later to become Lord Jackson of Burnley) in electrotechnics, J. R. Hicks in political economy and R. A. C. Oliver in education. P. M. S. Blackett (later Lord Blackett) had come only a year earlier to succeed W. L. Bragg. I had little to do professionally with Blackett apart from skirmishes on the Senate – physics and chemistry being the two dominant departments in the university we were careful to treat one another with respect on Senate. However I used, from time to time, to find myself a minor participant in the frequent wordy battles between Blackett, Polanyi, L. B. Namier the historian, and Wadsworth, the then editor of the *Manchester Guardian* on political and economic matters or on the nature of science (one of Polanyi's hobby-horses). Disagreement between Blackett and Polanyi on the social responsibility of scientists and the freedom of science was profound. Patrick was endlessly pressing left-wing socialism while Mishi would have none of it: Wadsworth and Namier were somewhere in the middle. I was not, as I have said, deeply involved but, even if I did not go quite as far as Mishi, my political views were nearer to his and distinctly to the right of the others. There were, of course, all sorts of currents within the university and with something approaching a constant state of war between Lang and Drummond (the professors of botany) on the one hand and H. Graham Cannon, our zoologist, who had the lowest opinion of their subject on the other, there were few dull moments.

One of the pleasantest features of life in the University of Manchester in my time was that most of the teaching staff lunched at common table in the Staff House each day and, as a result, one quickly came to know colleagues in subjects far removed from one's own. A wide circle of friends could thus be built up quite quickly and with much less effort than is the case in a university like Cambridge, where the staff is broken up into small groups on a college and departmental basis. It is, of course, also true that in Oxford and Cambridge the lack of any body corresponding to the Senate of a university like Manchester tends to make inter-professorial contacts more difficult. In Manchester, staff from the Manchester College of Technology (which, although the Faculty of Technology of the University, was located about a mile away near the city centre) also came from time to time to the Staff House. My colleague there, James Kenner, Professor of Technological Chemistry, used to lunch with us fairly regularly and it was thus that I came to know him well. Kenner, a much older man than I, had a fearsome reputation as a quarrelsome man and, indeed, when I accepted the Manchester job there were several people who warned me that I would get a rough passage from Kenner, who would not take kindly to seeing a mere stripling in the senior chemical chair. They were quite wrong. Kenner could be quarrelsome and difficult and he relished chemical polemics, but this was largely because he had become embittered through what he regarded – perhaps not without reason – as lack of recognition of his chemical contributions and lack of support in the College of Technology. This is not the place to discuss the problems of a strange and difficult man; to me he showed only kindness, and I am grateful to him for many helpful discussions on various aspects of organic chemistry.

When we went to Manchester in 1938 my wife and I rented a largish typical Manchester villa in Broadway, Withington, and found ourselves living directly opposite another very well known organic chemist, Arthur Lapworth – one of the great figures in the development of modern views on the mechanism of organic reactions. Before this I had met him only once and that for a few minutes when Robert Robinson, one of his friends and a great admirer, introduced me to him at a Chemical Society meeting. Lapworth was Professor and

Director of the Chemical Laboratories when Heilbron came to Manchester but was dogged by ill-health and finally resigned his chair in 1935. From 1938 until we moved to Wilmslow in 1941 I used to visit him each Sunday morning and chat about chemistry and chemists over a cup of coffee. These coffee mornings were kept under careful control by his wife, Kate Lapworth – a dragon-like lady who would storm in after about three-quarters of an hour, collect the coffee cups and unceremoniously send me about my business. I greatly enjoyed those talks with Arthur Lapworth and learned a lot from him. Mrs Lapworth made little attempt to conceal her view that I was something of a come-down as occupant of her husband's former chair. For one thing I was not a Fellow of the Royal Society when I took office and I remember vividly her reaction when, greatly to my surprise and delight, I was elected to the Fellowship in 1942. On being told by someone about the election she merely grunted and said 'And about time too!'

During my first year in Manchester we completed our studies on vitamin E and effected the total synthesis of α-tocopherol and its analogues. We were forestalled in the synthesis of the α-compound by Paul Karrer of Zurich who completed it a week or two before we did. The reason for this was rather absurd; to do the synthesis we needed the complex alcohol phytol which is a component of chlorophyll. Heilbron was the only man in England who was said to have any of it and he sent me a sample to carry out my synthesis. Unfortunately he made some mistake, and the material he sent me was not phytol. It took me a little time to find this out and get hold of some genuine phytol, (from Hoffmann La Roche) and while I was doing so Karrer completed his synthesis. Such things happen in research, but in the long run are not really important. We also pushed ahead with our work on *Cannabis* using column chromatography extensively in attempts to isolate the active principle or principles in a pure state. In this we failed, largely because our purification procedures, including preparative chromatography, as it was at that time, were too crude to make such isolation possible; indeed, something like twenty years were to elapse before it became possible for Israeli workers to take up the fractionation of *Cannabis* resin again using improved techniques, and succeed in isolating the active principle. However, when we set about the synthesis of cannabinol, an inactive constituent which had been first isolated and studied by Cahn some years before, our route took us *via* an intermediate tetrahydrocannabinol which showed powerful hashish-like action in rabbits. Our view at that time that the physiological effects of *Cannabis* were due primarily to another of the several possible isomeric tetrahydrocannabinols was vindicated much later, but we did little more work in this field since by that time (1940) the tempo of war work was increasing and we were forced to abandon the *Cannabis* research and never really went back to it. The synthesis of a variety of analogues of tetrahydrocannabinol was the subject of a good deal of work by Roger Adams and his school at the University of Illinois, they being still free of wartime restrictions. I had something of a contretemps with Adams on priorities in synthesis and publication in the *Cannabis* field at that time; with hindsight it now seems rather trivial, and it did not prevent us from becoming fast friends when we met after the war ended in 1945.

When Germany invaded Poland in 1939 Alison and I were on holiday at Ballantrae in Scotland. We at once drove back to Manchester where we got busy with blacking out the laboratory windows and doing the same job at our house in Withington where we also had to carry out the conversion of our cellar into a tolerable air-raid shelter. Our first child – a son – was born to our great joy on 11 November 1939 during the quiescent period known as the 'phony war' which lasted until the spring of 1940 when the Germans overran France. We then had to make a difficult decision, for many university people were sending wives and children to Canada or the United States to spare them from the intensive bombardment and possible invasion of Britain which everyone expected, and we had urgent invitations from the Bests in Canada and the Paulings in California. After much thought we decided that we should face as a family anything that was to come rather than separate, possibly for years, and perhaps for ever. In the same way we decided, after a few weeks' trial later on, when Alison and our son Sandy stayed with some friends in the Vale of Clywd, north Wales, that we would also eschew even that degree of separation. We never regretted our decision to stick together, although, when the heavy bombing of the Manchester area took place in the winter of 1940–41 and we were expecting our second child, we moved out of Manchester proper and in the summer of 1941 went to live in Wilmslow, some miles out to the south, whence I commuted daily to the university.

The outbreak of war soon brought other preoccupations and responsibilities. The chain of events leading to my appointment to Manchester had begun with the death of Sir Jocelyn Thorpe who was head of the chemistry department at Imperial College. He with Robert Robinson and Ian Heilbron made up the Dyestuffs Group Research Committee of I.C.I. Ltd, which held monthly meetings with the research chemists and management of the Dyestuffs Group at the Group's headquarters at Blackley in north Manchester. Since Thorpe's death the Committee had been one short. I.C.I. Dyestuffs Division (as it was now to be called) was beginning to awaken to the possibilities of synthetic drugs, and clearly this was likely to be a field which would undergo substantial development under wartime conditions. Whether or not it was due to my interest in and connections with that kind of activity I cannot say, but, at any rate, I was invited in the autumn of 1939 to join Heilbron and Robinson on the Group Committee now re-named the Dyestuffs Division Research Panel. To it as the pharmaceutical activity developed under Dr C. M. Scott, there were added the pharmacologists A. J. Clark and J. A. Gunn, and Warrington Yorke, an expert in tropical medicine; years later the pharmaceutical activities were hived off into a new Division, I.C.I. Pharmaceuticals, the chemical members of the D.D.R.P. continuing to act as advisers thereafter to both Divisions. This association with I.C.I. Dye-stuffs and Pharmaceuticals begun in 1939 continued formally for over twenty-five years; it remains today a happy informal association which has given me many friends over the years.

Very soon after the outbreak of war I found myself involved

with chemical defence research and development. There was at the time an unwarranted assumption that chemical warfare would be let loose on the civilian population and (in my view) a grossly exaggerated idea of the attendant dangers. As a result it was not surprising that there was a good deal of time wasted initially in investigating potential chemical warfare agents and in trying to devise new ones by completely hit and miss methods. To say that we were ill-prepared for chemical warfare in this country (apart from the provision of gas masks for civilians) would be to put it mildly, and it seemed that in the matter of considering new agents, or even the manufacture of known ones, virtually no progress had been made since the First World War. I soon found myself drafted, first as a member then as Chairman of the Chemical Committee which, under the Chemical Board of the Ministry of Supply, was responsible for development and production of chemical warfare agents. The research establishment mainly involved in this work was at Sutton Oak, St Helens, not very far from Manchester and manufacturing facilities were available through I.C.I. plants at Runcorn, Widnes and Blackley. All these are convenient to Manchester and so I remained stationed there throughout the war and was able to keep a reasonable measure of my own research going in the university in addition to work for the war effort. The Chemical Committee used to make occasional visits to the Sutton Oak establishment during the first couple of years of the war while the fear of chemical warfare was still very much to the fore. I recall much discussion about the virtues of the 'brass still' in the purification of mustard gas; as far as I could discover, however, its only virtue lay in the fact that, for one reason or another, and probably merely because of its availability at the time, a brass still had been incorporated in a plant that worked successfully in 1918. Now in 1940 a similar item, it seemed, had to be put into any new plant which was to be built!

The situation regarding arsenical sternutators which had been developed and, I believe, used by the Germans in the First World War was also rather ridiculous. After that war our chemical defence people had undertaken a study of the German manufacturing process of diphenyl chloroarsine by what was known as the double diazotisation process, starting from aniline and proceeding *via* a procedure well known in the dyestuffs industry called diazotisation modified so as to permit the introduction of arsenic into the molecule. The result of the study was recorded in a voluminous report purporting to show that it was theoretically(!) impossible to get more than about thirty per cent yield of product. This appeared to me to be such nonsense that I undertook to show that I could produce an effective process for manufacturing diphenyl chloroarsine by this method. I was, rather reluctantly, given permission to have a go, and with two research students and a dyestuffs chemist we borrowed from I.C.I. we not only demonstrated that we could get yields approaching the theoretical in the laboratory, but we went up to Blackley Works and there made the necessary arylarsonic acid successfully on a pilot plant producing five tons per week. This was a great experience for me and for the boys in the laboratory; we not only vindicated our criticism of the government report, but we learned quite a lot about the problems encountered in passing from a chemical experiment in the laboratory to large-scale manufacture. Subsequently we used our experience to develop a pilot plant for the production of the so-called 'nitrogen-mustard', but abandoned it when we had a disastrous explosion while chlorinating a sample of methylamine which, unknown to us, was contaminated with ammonia; we wrecked a laboratory, but by sheer luck no-one was injured. I need hardly add that our efforts in these areas contributed precisely nothing to the war effort since chemical weapons were never used.

However, my chemical defence commitments had their lighter moments. I recall being called upon to travel down to the Defence Research Establishment at Porton to watch a demonstration of a new chemical weapon for use against tanks. It must have been in 1941, because air-raids were heavy and frequent, tobacco was very scarce and, as petrol was equally hard to come by, I travelled down from Manchester by train through Bristol to Salisbury. As it happened Bristol had a big raid on the night my train was passing through and we had to lie stationary in the railway yards with bombs dropping uncomfortably close until the raid ended. We then trundled on through the placid countryside of south-west England and arrived at Salisbury around 8 a.m. where I was to breakfast before setting out in an army car for the demonstration on the open plain near Porton. Now, in those days I was quite a heavy cigarette smoker and the modest supply I had wheedled from my supplier in Manchester had long since gone and I was rather desperate; but, needless to say, I could find no-one in Salisbury who would supply me with any. So I breakfasted, trundled off to Porton, watched gloomily a rather unconvincing weapon demonstration and was taken to the local officers' mess for lunch. After having a wash I proceeded to the bar where – believe it or not – there was a white-coated barman who was not only serving drinks but also cigarettes! I hastened forward and rather timidly said 'Can I have some cigarettes?'

'What's your rank?' was the slightly unexpected reply.

'I am afraid I haven't got one,' I answered.

'Nonsense – everyone who comes here has a rank.'

'I'm sorry but I just don't have one.'

'Now that puts me in a spot,' said the barman, 'for orders about cigarettes in this camp are clear – twenty for officers and ten for other ranks. Tell me what exactly are you?'

Now I really wanted those cigarettes so I drew myself up and said 'I am the Professor of Chemistry at Manchester University.'

The barman contemplated me for about thirty seconds and then said 'I'll give you five.'

Since that day I have had few illusions about the importance of professors!

In wartime air-raids, fire was perhaps the greatest danger we had to contend with, and the university – and especially departments such as chemistry – took fire precautions very seriously. At the outbreak of war word was sent to all departments to examine all attics and ensure that they contained no inflammable materials. I well remember our venture into the attics in chemistry. We must have been the first entrants for many years, and we were somewhat taken aback to find in straw-lined open boxes a considerable number of sealed bulbs containing metal alkyls which dated from the last century and evidently belonged originally to Frankland (their discoverer). As these compounds are spontaneously inflammable they represented a considerable but entirely unsuspected fire risk even in peacetime! Another somewhat alarming discovery was made when we decided to look at our basement as well as our attics. Here we found several large bottles of mustard gas and a substantial amount of rather ominous looking decaying cordite; these we decided not to touch ourselves but to have dealt with by an army bomb-disposal unit.

Following this preliminary clean up, firewatching was put on an organised footing. In the chemistry department we had a rota system in which no distinction was made between teaching staff, technical staff and research students although it happened that most of the technical staff were engaged in firewatching or Home Guard duties in their home areas so that our departmental firewatching teams were made up for the most part of staff and research students and included about once or twice a week Ralph Gilson, F. S. Spring, at that time Lecturer in Organic Chemistry, and myself. These were rumbustious nights with every conceivable practical joke laid on and very little sleep for the participants even in the absence of raids. In my view the firewatching system was a huge success in a quite unexpected way. It brought into intimate contact research students and staff and I believe it made the Manchester chemical school of those days into a tightly knit group and set up relationships of mutual trust and respect that have endured to this day. I for one shall always cherish the experience of those nights on duty.

By the summer of 1941, when I moved with my family to our house in Pownall Park, Wilmslow, we were glad to have a good garden where we could grow vegetables and keep a few hens, for food was none too plentiful and at times not very good. We were especially fortunate because with the house we inherited a part-time gardener and odd job man, Harry Hardy (or 'Arry 'Ardy as he would have put it), an ageing native who had worked locally on the land all his days, and who lived in

a minute cottage nearby on the edge of Lindow Common.
Hardy had a smallholder friend, George Potts, on whom we
used to drop in of a Sunday morning, and through him we were
able to get a decent supply of fruit, of potatoes and other
vegetables, and even the odd chicken or duck. Petrol was, of
course, also difficult to come by and I recall when my wife was
in the maternity hospital a mile or two away at Prestbury for
the birth of our first daughter, Helen (on 13 July 1941), I used
to traverse the hot dusty road from Wilmslow to visit them on
a 'utility model' bicycle which Ralph Gilson had somehow or
other obtained for me from a little shop in one of the less
salubrious sections of Stockport Road near Ardwick.

One evening in the late winter of 1943 – I seem to remember
that it was in February or March – I was on firewatching duty
in the laboratory and had gone to do some writing in my room
when about 9 p.m. the telephone rang. I picked it up and heard
a voice say 'This is the Vice-Chancellor of the University of
Cambridge. The electors to the chair of biochemistry have just
met and it is their unanimous wish that I ask you to accept
the chair as successor to Sir Frederick Gowland Hopkins. We
wish to publish the name of Sir Frederick's successor on the
Senate House wall tomorrow and would like to have your
agreement.' To say that you could have knocked me over with
a feather would be something of an understatement. I knew,
of course, that Hopkins was retiring; but I not only knew that
I was by no means a biochemist as the academic biochemists
understood the term, but I also knew the person who was
generally expected to succeed Hopkins and his name was not
Todd. So I stammered 'Do you mean you would like me to give
you an answer to your question now?' 'Yes,' came the reply.
'Then Vice-Chancellor the only reply I can give you is no.'
There was a moment of silence at the other end and then the
Vice-Chancellor said 'Would you hold on for a moment while
I have a word with the electors.' I then heard sounds of a
discussion in the background and after a moment or two I had
in succession my father-in-law Sir Henry Dale, Sir Robert
Robinson and Sir Charles Harington on the line. The burden
of their song was 'Don't insult Cambridge by turning it down

flat – at least agree to come and have a look at it.' The upshot was that I agreed that I would go to Cambridge, look at the biochemistry department, and defer my answer until I had done so.

Accordingly, a week or so later I travelled down to Cambridge and stayed at Emmanuel with the Master (Dr Hele), himself a biochemist, and spent the following day looking at the position in the Sir William Dunn Department of Biochemistry. It quickly became clear to me that it was no place for me. It is difficult now to summarise my reactions without appearing to be unfair but briefly they were as follows:

(1) There was no real unity of purpose in the department. It was a series of little independent kingdoms sharing the departmental budget between them and the only gesture to unity was an almost sycophantic attitude to Hopkins on the part of the leaders in each of them.

(2) With the exceptions of Robin Hill and F. G. Hopkins himself the staff seemed to have little or no interest in the only aspects of biochemistry in which I had any expertise.

(3) The teaching courses were to my mind thoroughly inadequate as regards their chemical content and could produce no students who would fit into my type of work.

(4) I knew that I had a large number of people who wished to do research with me and on whom the progress of my work depended; there was simply no room to accommodate them in the Cambridge biochemical laboratories.

So, much to the indignation (but also I suspect to the relief) of the staff of the Cambridge department of biochemistry I refused the chair. When I met the Vice-Chancellor-elect and conveyed my decision to him, he told me he was not really surprised; he told me, however, that in the following year they were hoping to elect a successor to Sir William Pope who had died early in the war and had left the chair of organic chemistry and headship of the chemistry department – or to give it its proper title the University Chemical Laboratory – vacant. He realised that I would be unable to make any kind of commitment, but wanted to know whether this might be of more interest to me. I rather guardedly said that, subject to various conditions, it might be, and returned to Manchester.

This was not indeed the first time that I had heard of this possibility. Pope died in the autumn of 1939 not long after the outbreak of war and, shortly thereafter, Robert Robinson asked me whether I would be prepared to go to Cambridge if I were approached. I said I would not, partly because I considered it would be unfair to Manchester and to its Vice-Chancellor who had given me my chance, and partly because until the war was over I had plenty to do without trying to move to Cambridge. Robinson accepted my view readily enough, but told me not long afterwards that it had been agreed to hold the chair vacant until the war was over when he hoped I would think about it again. By 1943 the tide of war was turning so that a decision by Cambridge to proceed to an election in 1944 was understandable.

I now had to do some hard thinking for it was obvious that having rejected biochemistry I was going to be asked to take organic chemistry. The choice before me – Cambridge or Manchester – was by no means an easy one. I had in Manchester a large and thriving department with about thirty research workers in organic chemistry and the quality of our undergraduate input was at least as good as – and possibly better than – that of any other English provincial university. We were housed in old buildings but I had some reason to believe that the construction of a new chemistry department on an adjacent site figured in the university's development plans, although I did not know how high it stood in the priority list. I also had in Sir John Stopford a Vice-Chancellor whom I admired tremendously and who had helped me at every turn, and, in Michael Polanyi, a colleague whom I liked and with whom I was on the best of terms. Furthermore, my wife and I were very happy with both our social and geographical situation and were in no hurry to move. Against this, of course, we had to weigh the claims of Cambridge, taking a long view of the likely consequences of moving there. The Cambridge laboratories I had only seen rather fleetingly on the occasion of a British Association meeting in 1938, but my recollection was that they were rather old and badly equipped. As a school of organic chemistry Cambridge was virtually non-existent; its research effort was negligible in size, although the capacity of the laboratories (substantially occupied during the war by the chemistry departments of two London colleges – Queen Mary College and St Bartholomew's Hospital Medical School) was considerably greater than that of Manchester. The amount of actual or potential research accommodation was of importance, for I was quite well aware that the nature and success of our Manchester research would almost certainly lead to my being faced with a large influx of would-be research students both from elsewhere in Britain and from overseas, as soon as the war ended. It was also undeniable that, by long-standing tradition, Oxford and Cambridge had virtually first choice of applicants for university admission. My wife too had been up at Newnham College, Cambridge and both her father and her younger brother Robin were Trinity men; as a result we had many friends in Cambridge, and there was no doubt about the attractiveness of Cambridge as a place in which to live. I had also to consider the future of my lively Manchester group.

Within a month or two of my having turned down the offer of Hopkins' chair I was visited in Wilmslow by J. T. Saunders, Secretary General of the Faculties and the real head of the academic administration in Cambridge. He was sent up to talk to me, in strict confidence, about the general situation of chemistry in Cambridge, and to explore the kind of conditions I might wish to lay down if officially approached to take the organic chemistry chair. Some of the problems which beset chemistry in Cambridge I already knew and others I suspected. W. J. Pope had been head of a department which embraced not only organic but also physical and inorganic chemistry. As far as I knew – and this was broadly confirmed by Saunders – Pope had ruled his colleagues with a rod of iron and, during the last part of his life, had made matters worse by ceasing to take much interest in chemistry and becoming almost a recluse. There was a long history of internecine warfare between R. G. W. Norrish who was Professor of Physical Chemistry and E. K. Rideal who, although also running what was in effect a second school of physical chemistry, did so under the banner of Colloid Science. Furthermore, the chair intended for inorganic chemistry had been diverted into one entitled Theoretical Chemistry which was occupied by J. E. Lennard Jones whose interests were mathematical rather than chemical, and who had no contact at all with inorganic chemistry, important though that subject was (or should have been) in undergraduate teaching. Not surprisingly perhaps, Norrish was determined that he would not be at the beck and call of another Pope, while neither Lennard Jones nor Hamilton McCombie (Reader in Chemistry who was acting as a kind of caretaker of the organic laboratories) would contemplate being subservient to Norrish; I understood from Saunders that the department was being run (not very efficiently) by a committee of these three men, although Lennard Jones was absent as a temporary civil servant with the Ministry of Supply during most of the war. In the main chemical laboratory (physical chemistry was housed in a separate but adjacent building with a connecting passageway) things were at a low ebb. Pope had ceased to take any active interest in his subject, stereochemistry, several years before his death, and command of the laboratory had passed to W. H. Mills, a somewhat narrow stereochemist of a rather sour and, to young men at least, forbidding disposition who discouraged all but his own rather limited field of work. Mills had retired during the war leaving, as senior organic chemist, F. G. Mann, a university lecturer and Fellow of Trinity. Mann had come from London to Cambridge in 1919 to be research assistant to Pope, and had remained in Cambridge ever since. He certainly had a hard life under Pope and Mills and, by the time Mills left, he had more or less shut himself off with a few research students pursuing his own work, which was also stereochemical like that of his mentors; his interests lay largely in the stereochemistry of co-ordination compounds of metals on the one hand, and heterocyclic derivatives of phosphorus and arsenic on the other. Mann was a first-class chemist but much embittered by his experiences in Cambridge, and by the fact that his interests lay in areas that had become unfashionable and in some respects sterile. He did little to hold the department together, and the task of keeping it afloat had passed to the amiable, although chemically rather ineffective,

Hamilton McCombie. The latter had at least taken some action about inorganic chemistry by arranging that H. J. Emeleus, then a reader in Imperial College London, should come down each year and deliver a course of lectures to the Cambridge undergraduates.

Having heard what Saunders had to say I told him that I did not think a chemistry department could be run by a committee and that I could not accept such a proposition, but I agreed to his request that I should come to Cambridge, meet with the resident electors to the chair, and give them my considered views after looking over the laboratories.

In due course I was formally invited to go to Cambridge for this purpose, but meanwhile Alison and I debated again and again the question of moving, and my correspondence on the matter with Robert Robinson on chemical matters, and Alison's father on Cambridge and on diplomatic aspects, grew apace. Throughout all our discussion on Cambridge Sir Henry was a great help, for he was an immensely wise as well as a kindly man. The academic grapevine is, of course, very efficient, and by now the possibility that I might move to Cambridge was being widely, if not openly, discussed in Manchester and particularly, of course, in the chemistry department. I gathered that quite a few members of my research school were minded to accompany me to Cambridge if I went, and Ralph Gilson told me quite openly that if I moved to Cambridge without him he would leave university work altogether and take up some other career. While all this was flattering, it was also just a little alarming, since it looked as if there might be a mass exodus in the event of my deciding to move.

When I went to see the Cambridge electors I first had a good look at the University Chemical Laboratory in Pembroke Street with McCombie (whom I had come to know through his association with Chemical Defence) as my guide. When I saw it my heart sank; it was quite dreadful. To begin with, the main part of the building dated from 1886 and, unlike the Manchester laboratory, it had been poorly maintained; one block had been added after the First World War on the strength of a substantial endowment from the Anglo-Persian Oil Company; but the new building was built on the cheap according to ideas of design already out of date, while the university, in effect, channelled the rest of the endowment into general university funds. Even in this newest block, the laboratories were lit by gas although obviously gas brackets on the laboratory benches represented an appalling fire hazard. On enquiring how this had come about, I was given a fantastic explanation the gist of which was that determination of the end-point in volumetric analysis by titration was more accurate by gaslight than electric light! I could not help wondering whether greater significance might not have been attached to the fact that, according to my informants, since the Cambridge Gas Company had come into existence successive professors of chemistry in Cambridge had been members of its board of directors. The level of equipment was very low and what there was was mostly antiquated. There appeared to be no coherent organisation at all, administration being in the hands of a well-meaning and, within his limitations, able enough senior laboratory assistant, Charles Lister, who had joined the laboratory staff at the age of fourteen and was now coming towards retirement. The academic staff rode roughshod over Charlie and, as a result, the whole place consisted of small virtually autonomous units, each run for his own benefit by a member of staff. I suppose they had little choice, but it was clear enough that this system of operation led to gross waste and inefficiency; as a result of it, no-one had the faintest idea of the total amount, or even the whereabouts, of chemicals and apparatus in stock. At the time of my inspection only a small amount of space was occupied by F. G. Mann's stereochemical work and an equally small part by B. C. Saunders, who had temporarily given up his work on the mechanism of peroxidase action and was doing some first-class research on organophosphorus compounds for the Chemical Defence section of the Ministry of Supply. The remainder was occupied by (*a*) Queen Mary College, (*b*) St Bartholomew's Hospital Medical School, and (*c*) a small group carrying out work on the separation of uranium isotopes for the atomic bomb project.

I really was appalled by what I saw and at first felt I should withdraw at once. A little reflection indicated, however, that if I could get rid of the three groups of squatters mentioned above, have some money to change lighting and equip the laboratories to modern standards, and set up a proper administrative organisation, the place had great potential. Moreover, the school was so run down that I was unlikely to meet with any serious opposition when I started to put it in order. Having come to this conclusion I then confronted the Cambridge electors and the Secretary General of the Faculties. I had to begin by telling them there wasn't a great deal to attract a newcomer, the Cambridge school being currently one of the weakest in Britain and the University Chemical Laboratory a disgrace to any university. Indeed, I seem to remember suggesting to the chairman that given a stuffed crocodile to hang from the roof, the professor's private laboratory could be more appropriately located in the Museum of the History of Science, which was one of the projects being discussed in the university at that time. I must confess that the electors took it very calmly, and did not appear to disagree with me; they asked me under what conditions I would consider appointment and I gave them the following:

(1) I must have complete authority as head of department to reorganise and develop the University Chemical Laboratory. Whether physical chemistry remained with the rest of chemistry or hived off as a separate department was of no great importance to me at the moment. (I was almost certain, of course, that if Norrish were given the chance he would go for a separate department – as indeed he did. Whether this separation was wise is arguable, but, in practice, it dealt reasonably satisfactorily with the situation as it was in 1944.)

(2) Queen Mary College and St Bartholomew's must go back to London and the uranium work must be transferred to some other department or, better, to a government laboratory where it should have been in the first place.

(3) A post must be provided for A. R. Gilson to act as Laboratory Superintendent in charge of all non-academic affairs.

(4) Gas lighting should be abolished and replaced by electricity, benches provided with adequate power facilities, and money provided to equip the laboratories as quickly as possible to modern standards.

(5) I would need an undertaking that the university would give the highest priority to building a new University Chemical Laboratory on a fresh site as soon as possible after the war. I knew that similar priority had been asked for an extension to the University Engineering Laboratory, and I had to insist that a new chemical laboratory be regarded as of similar urgency.

(6) I would require a minimum of one academic staff position forthwith for my colleague Dr B. Lythgoe, who was then a Lecturer in Manchester and a key figure in my nucleotide coenzyme group.

Thereupon I returned to Manchester and awaited results.

There was naturally a little delay before the Board of Electors could meet and the university's attitude to my demands could be ascertained. But within a few weeks, after a further check by the Secretary General to be sure I had been properly understood, I was informed that all my conditions would be met and I was invited to take the chair. I accepted, although I remember wondering very much whether I had done the right thing. But the die was now cast, and things began to move quite quickly. I soon found myself in receipt of letters from heads of several colleges in Cambridge inviting me to accept a Professorial Fellowship. As I knew literally nothing about any of the colleges, except that they varied in size and I wanted, on the whole, to be attached to a medium-sized one I consulted my father-in-law. As a result I finally chose to go to Christ's because it was recommended to me as a smallish college with a decent reputation and one which had, in Charles Raven, a notable figure as Master. It is only fair to add that the presence of J. T. Saunders, the Secretary General, as a Fellow was an added attraction. Not only did I like Saunders – he and I became very good friends – but it seemed to me that, in battles which I might well have with the university authorities, the presence of the Secretary General as a Fellow of my own college could hardly be disadvantageous!

Since I had agreed to take up the Cambridge appointment on 1 October 1944, which was little more than six months ahead, we had to get busy at once in Manchester to prepare for the move. Ralph Gilson examined the Cambridge laboratories in some detail and, when he had recovered from the initial shock, set on foot their re-equipment and got under way the plans for removing the gas lighting and doing some minor structural alterations which were necessary. It was, of course, very difficult to get anything done in those days, since virtually everything was directed to the war effort and everyday civilian needs had to go largely unsatisfied even when, as in our case, the necessary money was available. But we were lucky, partly because of the private arrangements we had made with equipment manufacturers in Manchester, but, more particularly, because we were involved in the research effort on penicillin. The word penicillin was a real talisman at that time, and we used it to the full to get both government grants and – much more important – permits and priority for alterations to the laboratory. I then discovered that almost all the members of my personal research group in Manchester (other than those who had completed their course of research and had accepted positions in industry) wished to accompany me to Cambridge. This I found very touching because in some cases such a move would mean prolonging their Ph.D. course by a year, and for all of them it would be a great upheaval; I don't think I had really appreciated until then the loyalty and enthusiasm of the group and their attachment (however undeserved) to me personally. The research students concerned were: F. R. Atherton, J. Baddiley, A. Holland, G. A. Howard, H. T. Howard, R. Hull, G. W. Kenner, L. E. Lyons, D. H. Marrian, P. B. Russell, P. Sykes, A. Topham, W. S. Waring and N. Whittaker. In addition there were, of course, Dr B. Lythgoe and A. R. Gilson moving with me and Barbara Thornber. Barbara had taken her degree in 1943 and asked me then what she should do as she didn't think research would suit her – indeed, she confessed that her main reason for electing to do chemistry at Manchester had been simply that her school record in chemistry was rather better than in other subjects. She helped me out by taking some training in analytical work for some months and then doing micro-analysis during a period when our professional micro-analyst had gone down with tuberculosis. When she heard I was going to Cambridge she said she wished to go too – which was not surprising as she was a popular member of the department and much identified with the group of research students named above, several of whom were her contemporaries. The problem was – what kind of opening was there for her in Cambridge? I said I would need a secretary, and she said that, if I gave her three months to learn something at a secretarial school, she would take the job. I said 'Done!' and she not only got the essentials of the job in three months, but then came to Cambridge and became for seven years the best personal assistant and secretary I have ever had. H. T. Openshaw, whom I had appointed to a staff position in Manchester a few years before, also wished to come to Cambridge but, as there was no suitable staff position available there, he finally decided, to our mutual regret, to remain in Manchester.

I should mention at this point that the people named above including Openshaw, but with the exception of Holland (whose present whereabouts is unknown to me) and Waring, represent the core of a remarkable dining-club known as the Toddlers Club. The club was formed in 1971 at the instigation of Ralph Gilson and its members dine together annually; membership is limited to myself and those who either moved with me from Manchester to Cambridge, or had arranged to work with me at this time. (These latter were A. W. Johnson* from Imperial College *via* wartime work in I.C.I., C. H. Hassall from Dunedin, New Zealand and J. Davoll, a Cambridge undergraduate.) Although now widely scattered in the country and abroad, the Toddlers (eighteen in all) maintain close contact, and all gather in Cambridge for the annual dinner in May. So far one member (G. W. Kenner) has been lost by death, and only one has been added – Herchel Smith, who did his undergraduate work in Cambridge and, like Davoll, joined my group there.

I left Manchester after much hesitation and with much regret; the University of Manchester will always retain a very firm place in my affections. After all I was a very raw youngster when I was appointed, and, looking back, I think the university, and especially its Vice-Chancellor, were taking a considerable risk when they chose me in preference to at least

* Deceased 5 December 1982.

two other well-known chemists (both my seniors) whom I later discovered they had also interviewed. As it turned out, however, and thanks in no small measure to the help I had from everyone in authority and the loyalty and friendship of my staff and students, I think I justified their faith in me. I learned a tremendous amount during those six years – in fact I grew up. I learned how to organise and administer a department without getting bogged down to the extent of allowing administration to interfere with research; I learned too how to choose a research topic and, having done so, to pursue it without allowing myself to be deflected into the interesting little side alleys which always turn up, and which have been the undoing of so many. In a way research is like golf – you must keep your head down and your eye on the ball!

Sometimes people say it is a pity my stay in Manchester was effectively during the war years. No doubt there would have been a wider social life had there been no war, but, against that, I doubt whether the intimacy between me and my students would have developed to anything like the same extent in peacetime. It is also to be remembered that Alison and I were recently married and that our first two children were born during our stay in Manchester. In the early days of a family, I rather think it would take more than a war such as we had to blunt in any way the happiness we enjoyed.

The move to Cambridge was successfully accomplished with little interruption in our researches, although I think the university was rather taken aback by the great invasion. By the time we moved, the uranium workers and St Bartholomew's had gone and Queen Mary College was due to leave shortly. We had got in a good supply of the equipment we needed, and, although we had not got rid of the gas lighting, its removal and replacement were imminent. Before it was replaced, however, we had several fires in the laboratory. One I remember was alarming, although it did not do a great deal of damage. A member of the research group – I think it was Norman Whittaker – held up a large flask of near-boiling petroleum ether to the gas light over his bench to see whether all the solid material in it had dissolved; the resulting conflagration was quite spectacular.

Our family removal to Cambridge also had its lighter side. I was able to purchase a house in Barrow Road, but could not get possession of it until the war was actually over. To bridge the gap, I was able to rent a furnished house in Sedley Taylor Road belonging to Sir John Cockcroft who was, of course, involved elsewhere in the wartime atomic bomb project; the house was to be vacated by its then tenants, the Rev. J. Boys-Smith and his wife, in September 1944. Now at this time, owing to acute housing shortages, a house left empty for more than twenty-four hours was liable to be seized by the local authority or by squatters, so we had to make rather unusual arrangements for the move. Using our penicillin talisman we hired a lorry in Manchester which would take a substantial part of our laboratory equipment (as well as the personal effects of most of the migrating research school and the bits and pieces of household equipment we needed urgently), and arranged that it should go to Cambridge on the exact day when Boys-Smith was moving out. With the lorry went three of my research group, who camped out in the Sedley Taylor Road house until I and my family came down a day or two later. My own most vivid memory of that move is of the trip I made back to Wilmslow a day or two later to fetch the family cat, a black and white animal of dubious ancestry who gloried in the name Sir Samuel Hall. Sir Samuel did not approve of either the basket in which he was confined or the journey, and said so vigorously; we had an uproarious half-hour while changing trains at Rugby, while Sam competed vocally with about a hundred cheerful Italian prisoners of war who were being transferred from one camp to another. Overall, however, the move to Cambridge went pretty smoothly, and by the end of September 1944 we had settled in both at home and in the old University Chemical Laboratory in Pembroke Street.

In the laboratory Gilson and I began the process of reorganisation and revitalisation forthwith. We met with very little opposition, and even F. G. Mann who, perhaps not unnaturally, felt he had greater claims to the Cambridge chair than I had, gradually, if reluctantly, came round, after some rather tiresome behaviour, to the view that our reorganisation made life easier and better for everyone, including himself. From the start we had the full support of B. C. Saunders and of Hamilton McCombie, who, until his retirement not long afterwards, was a tower of strength and a mine of information about Cambridge in general and its chemistry school in particular. We began by tackling the library and introducing a centralised system for the purchase and issue of chemicals and equipment. My predecessor had used the departmental library as his private room; I gathered that access had been possible only through the goodwill of Miss Stoakley (the daughter of the former laboratory steward) who acted as a kind of secretary, although she had no shorthand and was a rather indifferent typist. This had two results – firstly that the library was inadequately used, and, secondly, that there was considerable uncertainty about the ownership of some of the contents (I discovered, to my surprise, that many of the books bought by the laboratory were stamped with my predecessor's book-plate). This uncertainty of ownership may explain why, following Pope's death and before I came to Cambridge, Pope's executors had sold a considerable section of the library, presumably in all good faith, including a complete run of Liebig's *Annalen der Chemie* to Boots Ltd, and added the proceeds to the estate. Fortunately, I was able to restore the position by getting some journals on permanent loan from the University Library, so that we did not suffer too much. But I remember well Charlie Lister's astonishment when I told him I did not wish to have my book-plate put in any books purchased by the laboratory! In setting up a central store (and incidentally reorganising the whole office administration) we called in everything lying around in private stores of present staff members and in the forgotten stores of those who had gone before, and combined them with a weird collection of stuff (including a dozen Lewis guns fitted with cameras and several gross of milk bottles) which was housed in a basement room called 'the store'; we did meet a little resistance from the owners of private stores, but they soon came round when they found how much better the new system was.

During the first year or so we were without two senior members of the Cambridge staff, F. B. Kipping and P. Maitland, who were serving with the armed forces. When they returned they both contributed a lot to our progress. Kipping I put in charge of all academic administration, and this he did excellently until his untimely death some fifteen years later. During that first year I was also able to persuade H. J. Emeleus to leave Imperial College and join us as Reader in Inorganic Chemistry. Emeleus was the outstanding inorganic chemist in the country, and we were extremely fortunate to get him; his Readership was soon converted into a Professorship, and inorganic chemistry was back on the map in Cambridge to everyone's advantage. We had a little trouble at first getting hold of some laboratory space for Emeleus' research. What would normally have been his area had gradually been taken over by U. R. Evans and his corrosion group (which really should have been in metallurgy and which, indeed, I transferred to that department when it was reorganised under Professor Austin). Corrosion work kept on acquiring space, because its progress appeared to demand that a continually growing series of laboratory benches should be covered with little glass beakers containing a variety of fluids into which pieces of metal had been placed, and which were simply left to stand there for months, or even, in some cases, years. The problem was solved when I pointed out to Evans that there was no need to cover all the benches with his beakers, and that these might be stacked somewhere in much less space. This apparently novel idea was accepted, and at once a substantial amount of laboratory space became available for Emeleus!

When I went to Cambridge I found that the university was run very differently from Manchester – and, I would guess, less efficiently. From my standpoint as head of a big scientific department, the Cambridge system of making a block grant which was paid into a departmental bank account each year was a very useful one, for it meant that the distribution of expenditure over various items in the budget could be varied a good deal, provided always that the auditors gave a satisfactory report on the accounts for transmission to the university's Financial Board. We did very well out of it in the early years, since our reorganisation soon revealed that the departmental budget was really much more adequate than had previously been believed, and the university was always prepared to listen sympathetically to us when we had a good case for increasing it. The university's contribution, together with generous grants placed at my disposal (especially by the Rockefeller Foundation and Hoffmann La Roche) in support of my researches, greatly facilitated the rapid build-up of my research school in Cambridge. Nevertheless, it seemed to me that the university's control was too slack during my early years in Cambridge; I was probably right in this, because it was tightened up very considerably later on. Although it, too, was changed within a few years, the system under which the university teaching staff was paid struck me as quite extraordinary. The stipend of a university lecturer was quite derisory (I think it was £150 p.a.) and included payment for a few hours of teaching. Undergraduate teaching over and above this was paid at an hourly rate. If the lecturer was a Fellow of a college he would receive in addition a college stipend plus payment for teaching or other work in college; in this way he could make quite a reasonable living. If he were not a Fellow, however, he received from the university a modest 'Fellowship Allowance', but if he were to get a living wage he had to do a lot of extra teaching. I well remember each year sitting down with a list of my staff members and, having first worked out what would be a reasonable salary for each individual, calculating the almost astronomical number of hours extra teaching (largely fictitious) they must do to get it and solemnly entering it in the official departmental return to the university.

After just about a year in Cambridge the war came to an end and the pace of our life quickened. We became finally settled as a family by moving into our own permanent home in Barrow Road in the late spring of 1945 and in the summer of the following year on 25 June our daughter Hilary was born. But even before then the influx of overseas students – soon to become a flood – had begun and I found myself becoming increasingly involved in a variety of activities in Cambridge, London and overseas. Throughout all of these my researches were continuing, and indeed expanding, as a result of the increasing number of young chemists seeking to join our school. It is not easy to incorporate the progress of research in a narrative of one's career, and I think it will be simplest if, at this point, I set down a brief outline of the development of my major research interests over the years in a separate chapter.

5

*Chance and design in research. The road to
vitamins, coenzymes and DNA*

———

During the wartime period in Manchester I was able, despite
other preoccupations (and the need to attend far too many
committees which multiplied alarmingly in the university), to
make, with my colleagues, surprisingly good progress with
research aimed towards nucleotide coenzymes. Thus, we devised new methods of phosphorylation and began to understand the chemistry of the organic phosphates and poly-
phosphates, and we produced new methods of synthesis, not
merely for pyrimidines and purines, but for their glycosides (the
nucleosides). Alongside research, however, we were making
some notable improvements in laboratory equipment. Ralph
Gilson was quite remarkable for his appreciation of laboratory
needs, and for his skill in the design of new equipment to meet
them. In the exercise of this talent he had my full support, and
we turned it very much to our advantage in the following way.
Ralph would design and build a prototype; he would then offer
the design to one of the scientific equipment firms which would
be allowed to take up and market it, with the proviso that it
first supplied us with what we needed, and that, thereafter,
when we required a particular item we would have priority
over other customers. This worked excellently, and so we
developed new evaporators, shakers, drying pistols, etc. and,
in due course, also electromagnetically stirred autoclaves for
pressure hydrogenation and other reactions. Most of these
came into general laboratory use in due course, but we were
normally first in the field and so, even during the war years,

we became perhaps the best equipped chemical laboratory in Britain.

In addition to work for the Chemical Defence Research Department we had a number of other projects in our laboratories in Manchester which were regarded as being 'of national importance'. These included work on antimalarial and other drugs, but one of the most interesting was an attempt to isolate and identify the so-called 'hatching factor' for the nematode *Heterodera rostochiensis* (the potato eelworm). This parasite attaches itself to the growing roots of the potato plant and causes the disease known as potato sickness; it is found all over Europe, and in Britain alone is responsible for losses amounting to millions of pounds annually. The fertilised female worm remains attached to the potato root, its body swells up and, with the death of the worm, the posterior portion drops off into the soil as a leathery cyst containing some hundreds of larvae. The cysts lie dormant but viable in the soil until potatoes are planted again in their neighbourhood, when the larvae emerge and attack the potato rootlets. The reason for this hatching is that the growing potato roots secrete into the soil a substance (or substances) which specifically triggers the hatching of the encysted worms. My attention was drawn to this fascinating problem by Professors Leiper and Raistrick of London, who had made some preliminary examination of the problem to ascertain its feasibility; they had found that the mysterious hatching factor was produced by other solanaceous plants so that tomato plants could be used as a much more convenient source than potatoes, and had devised a rather crude but apparently effective biological assay. They decided to abandon the project when war broke out and I was invited to take it up. This I did and, with the help of Dr C. T. Calam who had been doing the preliminary work in London with Raistrick, I got the project under way in Manchester. We failed to isolate the hatching factor in Manchester and the same lack of success attended further efforts I made later in Cambridge with my colleague A. W. Johnson. We were, however, able to establish that we were dealing with a relatively small molecule or molecules (< 1000 in molecular weight and probably around

500) containing an unsaturated lactone group, but we could not isolate it. True, it was not very stable, but the real reason for our failure was twofold. Firstly, fractionation procedures such as chromatography and countercurrent distribution were insufficiently advanced, and, secondly, our biological test reached its limit when we were dealing with fractions active in dilutions of about 1 in 10^6, and which were evidently still heterogeneous. It is perhaps only fair to say that our isolation work in Manchester was brought to a standstill in 1944 when the greenhouses at Cheshunt where we grew our tomato plants were demolished by a flying bomb. Long before then we had found the physical separation from our tomato cultivation rather tiresome, and we did indeed make one abortive effort to alter it by cultivating another easily grown solanaceous plant – the black nightshade – on a piece of ground owned by the university in Fallowfield. This experiment had to be brought unobtrusively to an end about a year after we started it, when black nightshade, to the concern of local residents, began to appear in gardens all over south Manchester! Our work on the eelworm problem was thus rather disappointing, although we did synthesise a few compounds on the basis of what we knew about it, and some of these had weak hatching activity. It is, however, some consolation that, even today, the factor remains unisolated, and that the efforts of quite a number of investigators in several countries have, in essence, only confirmed what we established about it many years ago.

The potato eelworm hatching factor is, of course, only one example of the many fascinating problems presented by the phenomenon of obligate parasitism, and analogous substances influencing the germination of a number of parasitic plants also occur. My colleague, Dr R. Brown, who was a lecturer in botany in Manchester and subsequently became professor of that subject in Edinburgh, was interested in, and drew my attention to, the probable existence of specific germination factors for *Orobanche* and *Striga* species secreted by the host plants upon whose roots they battened. We did have a look at these problems later on in Cambridge, but again had little success. The first breakthrough in this general field of research did not come until 1971, when workers in the United States succeeded in isolating a small amount of the *Striga* germination factor to which they gave the name strigol and established its structure by X-ray crystallographic analysis. Interestingly enough, its molecule has some of the structural features which we suspect occur in the eelworm hatching factor.

Another piece of work which led to no useful result (as far as I was concerned at least) was that which we carried out on blood anticoagulants. During my first year in Manchester I was asked by my friend C. H. Best of Toronto if I would have a look at heparin, which the Canadians were producing at their Connaught Laboratories. He sent over one of his technicians (Arthur Charles) with a modest supply of material so that we could do some work on it, but we had no time to do much with it before war broke out and Charles had to return to Canada. I would probably have paid no further attention to the problem, since I really was not very interested in carbohydrate chemistry, had it not been for the panic about spies and Nazi sympathisers which occurred following the German invasion of France. All German subjects, including many Jewish refugees, were suddenly rounded up and interned (quite arbitrarily as far as I could see). One victim of this was my valued colleague and friend Dr Anni Jacob. She was arrested, put for a time in Holloway Prison in London, and then moved to an internment camp at Port Erin, Isle of Man. This was a tragic waste of scientific talent at a time when this country needed all it could find, and was equally tragic from a human point of view, since Anni was one of those Germans who, although without any Jewish ancestry, nevertheless elected to leave her country rather than live there under Hitler. I had a long battle with the Home Office before I finally obtained her release, but, between times, in an effort to relieve the boredom of her stay at Port Erin, I arranged with the Director of the Marine Biological Station there that she could collect and process various types of seaweed, and, in collaboration with me in Manchester, look for possible blood anticoagulants in them (it was known that the Irish seaweed product carragheen had anticoagulant properties). Although we carried on some of this work for a short time after Anni was released, it really did not lead us anywhere and we abandoned it; but it probably helped her to retain her sanity during her internment. While we were still in Manchester Anni married Dr Juan Madinaveitia, who had been associated with me in Edinburgh, London and Manchester, and who was himself a refugee from the Franco regime in Spain. Madinaveitia subsequently joined I.C.I. (Pharmaceuticals) Ltd and he and Anni settled down happily in Cheshire and brought up a family.

Needless to say, I took part in the wartime Anglo-American cooperative research project on penicillin, but I was not involved in the early stages and our efforts beginning in, I think, 1943, were on a relatively modest scale. Looking back at it now, I find it rather amusing that, by showing that penicillin readily formed a sulphoxide, we did indeed establish that it was a true lactam; furthermore, our sulphoxide has, in recent years, sprung into prominence as a starting material for much synthetic work on β-lactam antibiotics in general. I confess that I got little pleasure from our penicillin work, and that, I believe, for two reasons. In the first place I found the constant stream of research reports from all participants in the cooperative venture, amounting almost to a flood, very distracting, and, indeed, counterproductive in that they hampered the free development of my own ideas. Secondly – and this applied in variable degree to all our war work – I have always found it difficult to do good research unless the subject is one in which I have a strong personal interest. I think this latter point applies to many academic research scientists, and that is why they are usually not very efficient in industrial contract work. Indeed, my advice to an industrial firm with a research problem which it wishes to solve expeditiously, is to do it within the firm, taking external advice as appropriate but not to contract it out to a university. I know, of course, that in recent years several industrially oriented units for contract research have been set up in a number of universities; but these are quite different in outlook and staffing from normal university departments, and have had variable fortunes. I still think that the proper place for industrial research is in industry.

I have already explained how chance in the shape of George Barger introduced me to the vitamin field; from that introduction grew an interest in vitamins, and especially in the chemical reasons for their importance. These interests are at the heart of what most people would probably regard as my main contributions to research – the chemistry of vitamins, of nucleosides, nucleotides, coenzymes and nucleic acids. Such a view is entirely reasonable, since there can be little doubt that our nucleotide work and the establishment of the chemical structure of nucleic acids form the base on which molecular biology and modern genetics have developed in such spectacular fashion during the past quarter of a century. Yet, in addition, I have always had a deep interest in natural colouring matters – an interest triggered by my association with Robert Robinson in research on the beautiful red and blue colouring matters of flowers known by the generic name of anthocyanins. As a result, I have, during my career, done a good deal of work on natural colouring matters and especially on those remarkable pigments found in aphids, those well-known sucking insects which are the bane of many gardeners' lives; I shall recount something of that research later.

In Manchester we had laid the ground for our foray into the field of nucleotide coenzymes. We had developed new methods for nucleoside synthesis and for phosphorylation of nucleosides with dibenzyl phosphorochloridate, and had discovered a new method for phosphorylating amines using diesters of phosphorous acid and polyhalogen compounds. (This latter reaction was, in fact, accidentally discovered by F. R. Atherton when he tried to remove acid impurities from a solution of dibenzyl hydrogen phosphite in carbon tetrachloride by shaking it with ammonia; the whole mixture set to a semi-solid mass of dibenzyl phosphoramidate.) I need not elaborate in detail, but during the first few years in Cambridge we had effected the first of our coenzyme syntheses – that of adenosine triphosphate (ATP), the substance which is involved in phosphate transfer and acts as the necessary reservoir of energy for muscular activity in animals. We also not only settled the structure of the known natural nucleosides and nucleotides by synthesis, but established their stereochemical configuration as well.

Already in 1938 when I started work in this field I was of the opinion that nucleic acids might be involved in the transmission of hereditary characteristics as had been suggested by the earlier work of Griffith on pneumococcal transformation; Avery's demonstration in 1944 that deoxyribonucleic acid (DNA) was the transforming factor seemed to me to settle the issue. Curiously enough Avery's work – so beautiful and, to me, so convincing – did not convince everybody. In the summer of 1946 I attended a symposium on nucleic acids held in Cambridge by the Society of Experimental Biology to which I agreed to contribute a paper on 'The structure and synthesis of nucleotides'. At that symposium I remember a violent argument between E. Stedman, who stoutly maintained that histones and not nucleic acids were the carriers of hereditary characteristics, and a number of others and notably Caspersson who, like me (and with better evidence), was a proponent of nucleic acid. I think it was really at that meeting that I first met and talked with the main operators in the biology and biochemistry of nucleic acids, and heard from Astbury about his X-ray studies. My interest was aroused, and I began – again for the first time – seriously to consider the chemical structural problems presented by the two types of nucleic acid – ribonucleic (RNA) and deoxyribonucleic (DNA). The time was in any case propitious, because we were just on the point of completing our first ATP synthesis and we now knew sufficient about phosphates and their behaviour to make nucleic acids, chemically speaking, a bit less daunting than they were to most people at that time. So we began to think about the problem a bit, and to study the behaviour of simple nucleotides alongside our coenzyme work.

I was, of course, familiar with most of the chemical literature on nucleic acids and their component nucleotides. As a result of his work extending over many years, P. A. Levene had substantially clarified the structure of the simple nucleotides and nucleosides which can be obtained by hydrolysis, and deduced correctly that the nucleic acids were made up of nucleotides linked together in some way through phosphate residues. But he had, based on analytical methods we now know to have been inaccurate, reckoned that only two nucleic acids existed – one from plants (RNA) and one from animals (DNA) and that each was composed of four nucleotides present in equal amounts. More unfortunate still, he supported the idea that the nucleic acids might be simply tetranucleotides which formed colloidal aggregates in solution. From the moment I read his claims and views I found myself in total disagreement. For one thing, there was already some evidence to suggest molecular weights of 500 000 or more for DNA, and, in any case, its general properties suggested strongly that it was a macromolecular substance like protein or one of the polymeric materials studied by Staudinger, and even then appearing in commerce in the form of synthetic rubber and synthetic fibres. Nor could I accept the idea that we were dealing with polymerised tetranucleotide units; on the evidence available I had doubts about the constant composition of both types of nucleic acids and the claimed existence of only two acids. These views were, of course, fully confirmed during the years that followed, but belief in the so-called 'tetranucleotide hypothesis' was, in my view, largely responsible for the slowness with which biochemists and biologists came to realise the importance of the nucleic acids in hereditary transmission. I have some-times wondered whether the ready acceptance of the tetra-nucleotide hypothesis by many biochemists was not, perhaps, due to their belief that proteins with their manifold properties would be found to be responsible for all life processes, and they accordingly felt no need to look any further!

I have referred to the daunting nature of chemical studies in the nucleotide/nucleic acid field. At the time of which I am writing, and, indeed, throughout virtually all the work on nucleic acid components – and even on the basic purines and pyrimidines let alone their phosphorylated derivatives – poor solubility in organic solvents, difficulty of separation and purification, and lack of proper melting points or other reliable criteria of purity, made life very difficult for the chemist, and led to a lot of confusion and, indeed, errors in the literature. In those days we had only the early forms of chromatography available, and, apart from ultraviolet spectroscopy, and in the late stages of our work some minor applications of the newly developing technique of infra-red spectroscopy, we had to rely on the traditional methods of the organic chemist developed many years before for compounds quite different in type and in physical properties. Even in our own work we were led by such deficiencies into errors in the identification of simple nucleotides, at one point, by accepting the wholly invalid 'proof' of the structure of benzylidene-adenosine due to Gulland, and had to do quite a bit of work before we realised the true position. As we went along several observations began to loom large in our thoughts. For one thing, we were struck by the astonishing difference in the ease with which RNA underwent hydrolysis as compared with DNA and probably related to this the ease with which one could purify deoxyribonucleotides as compared with ribonucleotides. To get a really pure specimen of yeast adenylic acid (adenosine 3'-phosphate) was a very difficult task indeed, since recrystallisation seemed often to operate in the reverse direction; this should have suggested phosphate migration to us as a possible reason, but I am afraid it did not immediately do so. What was, of course, clear, was that the difference between the stability of the two nucleic acids lay in the sugar portion of the molecule, and this also suggested to me that, while the stable DNA was involved in hereditary transmission, RNA probably had some quite different function in nature, where its impermanence would be an advantage. As far as we were concerned the final breakthrough came in 1949 when Waldo Cohn at Oak Ridge, Tennessee, applied ion-exchange chromatography to alkaline hydrolysates of yeast ribonucleic acid, and isolated, not just the nucleoside 3'-phosphates hitherto regarded on the basis of studies by Levene and others as the sole products, but also the 2'-phosphates. In Cambridge my colleague D. M. Brown drew my attention to some earlier and virtually unnoticed work by

Fonö on the hydrolysis of glycerophosphates, and suddenly the whole jigsaw fell into place, and we could explain all the hitherto puzzling facts about RNA and understand fully the differences in chemical behaviour between RNA and DNA. Moreover, we were able to establish conclusively by synthetic studies the reality of the alkaline degradation of RNA, first to four cyclic nucleotides and thence to an equilibrium mixture of the 2'- and 3'-phosphates, which undergo interconversion in an acid medium, and we were able to explain the nature of ribonuclease action and its significance for RNA structure. Having done this, Dan Brown and I were able to put forward definitive structures for the two types of nucleic acid as unbranched 3:5-linked polynucleotides in 1951 at the 75th Anniversary Meeting of the American Chemical Society (our detailed paper did not actually appear in print in the *Journal of the Chemical Society* until January 1952).

It was now clear that both types of nucleic acids were unbranched linear 3':5'-linked polynucleotides, and that individual members differed from one another in molecular size and in the sequence of nucleotides present in them. Accordingly, methods both for synthesis and for sequence determination would in due course become important. We did indeed carry out some work on stepwise degradation of polynucleotides by chemical means, but it quickly became clear that such methods would be tedious in the extreme, and probably inaccurate. It seemed to me that little real progress was likely to be made until enzymes could be found which would chop up polynucleotide chains in a manner analogous to the specific enzymes which were known to attack polypeptides, and which were being used with such success in the case of insulin and other proteins by F. Sanger. Very few nucleotidases of such a nature were known at this time, and it seemed to me wisest to leave the further development of sequence determination to men like Fred Sanger, or to one of the numerous young men who were going out from my laboratory to build up new research groups in the nucleotide field in many countries. On the side of synthesis A. M. Michelson and I showed that the synthesis of oligonucleotides was quite feasible by synthesising dithymidine-3′:5′-dinucleotide. I confess that I have never been much attracted to the kind of repetitive procedures involved in synthesising either polynucleotides or polypeptides, so I left further developments in synthesis to my 'offspring' including such men as A. M. Michelson, F. Cramer, H. G. Khorana and C. B. Reese. The spectactular results which later emerged culminating in Khorana's synthesis of a gene are well known. There still remain serious problems to be solved, however, even in the oligonucleotide field; especially is this true of the ribonucleotides, where the presence of *cis*-vicinal hydroxyl groups in the ribose residue presents a difficult problem for the synthetic chemist.

In the course of our studies on methods for the phosphorylation of nucleosides (i.e. nucleotide synthesis) we developed and used as the method of choice what has come to be known as the 'triester procedure', i.e. one in which one uses an active diester of, for example, phosphorochloridic acid as phosphorylating agent and subsequently removes selectively one or two ester groups from the resulting triester to give either a di- or a monoester of phosphoric acid. Although demonstrated by us to give good results in synthesising a dinucleotide, it appeared for some years to be superseded by other quicker but less selective routes based on direct diester formation. I find it sometimes a little difficult not to say 'I told you so!' when I see the triester procedure now being belatedly adopted for oligonucleotide syntheses as the method of choice where selectivity as well as yield is of importance.

The double helical conformation of DNA was advanced by Watson and Crick about two years after our clarification of the chemical structure which made it possible. I and my colleagues played no part in the development of the Watson–Crick model, largely because our interests at the time were essentially chemical and we really gave little thought to the physical conformation of the polynucleotide molecules in nature. A secondary reason may have been the almost total lack of contact between physics and chemistry in Cambridge – a lack of contact which is all too common in universities. It is true that before Watson and Crick were allowed to publish their paper Sir Lawrence Bragg, who was head of the Cavendish Laboratory, insisted that D. M. Brown and I should approve their model (which we did!). The reason for this insistence (which is mentioned in Watson's book *The Double Helix* but with no explanation) lay in the fact that only a year or two earlier Pauling had published the α-helical structure for a protein. Pauling sent copies of his manuscript both to Bragg and myself, and I well remember Bragg coming over to see me in the chemical laboratory (for the first time since my arrival in Cambridge) and asking me how Pauling could have chosen the α-helix from among three structures all equally possible on the basis of X-ray evidence, and all of which he (Bragg) had indicated in a paper with Perutz and Kendrew. He was quite shattered when I pointed out that any competent organic chemist, given the X-ray evidence, would unhesitatingly have chosen the α-helix. It was a direct consequence of this that he decided that no nucleic acid structure based on X-ray evidence would go out from his laboratory without it first being approved by me!

When I saw the Watson–Crick model that day in their laboratory, I at once recognised that, by a brilliant imaginative jump, they had not only solved the basic problem of a self-replicating molecule, but had thereby opened the way to a new world in genetics. Maybe it is a pity that the physicists and the chemists were not closer at that time, but, even if they had been, we might at best have enabled the physicists to make the imaginative jump a year or so earlier, but probably not much more. Arising from our synthetic studies on simple nucleotides, I had long since learned that the nucleosides were effectively flat, and their stereochemistry indicated that, linked together by phosphate residues, they must form some kind of helical structure. We also knew, from X-ray studies of some of our materials by W. Cochrane and his group, that nucleosides and, indeed, their parent pyrimidines and purines were strongly hydrogen bonded. I recall telling Astbury of these views as early as 1947. I knew, of course, that the DNA molecule must contain some kind of code if it were to transmit hereditary characteristics, but, save in a very desultory way,

I never considered the matter very seriously. Thus, although well aware of Chargaff's analytical findings in 1950 and 1951, I never gave any serious thought to their possible significance as part of a physical arrangement of DNA which could provide the basis of the genetic code. All of which is just an illustration of the way in which scientists are very often blind to matters which happen to lie outside their own specific field of interest.

A further striking example of this last point is to be found in our work on organic phosphates, a facet of our nucleotide coenzyme studies. One of the problems we had to face quite early in our work aimed at coenzymes, most of which were unsymmetrical pyro- or triphosphates, was that the initial phosphorylation of a nucleoside by our normal method (using dibenzyl phosphorochloridate) gave rise to a triester, from which one esterifying group had to be selectively removed under very mild conditions which would not damage other parts of the molecule. One of our most successful devices for this purpose was to make use of (*a*) the electrophilic character of the CH_2-grouping in the benzyl residue and (*b*) the fact that diesters of phosphoric acid are strong acids, i.e. they have very stable anions. Thus, when a triester of phosphoric acid containing a benzyl (or, for that matter, an allyl) group is treated with a nucleophile such as a tertiary base, or an anion such as chloride or iodide, the nucleophile attaches itself to the benzyl or allyl group liberating the anion of a diester of phosphoric acid. The more powerful the nucleophile and the stronger the diester of phosphoric acid produced in the reaction the better it goes. Now, of course, ethylenic compounds are nucleophilic, but rather weakly so, and accordingly were not of any value to us in the nucleotide work. But, around 1952, my colleague, F. R. Atherton (now with Roche Products Ltd and one of the world's experts on organic phosphates) decided to explore their use in this reaction. In the course of his work he found that geranyl diethyl phosphate was quite stable, but geranyl diphenyl phosphate was unstable and underwent cyclisation to give a mixture of cyclic terpenes, diphenyl phosphoric acid (a very strong acid) being expelled. What had happened was, of course, that the isolated double bond in the geranyl residue was sufficiently nucleophilic to attack the allylic carbon intramolecularly. This was quite interesting, and was in accord with expectations. What we failed to realise was that he had, in fact, discovered the way in which nature makes carbon–carbon bonds! Our interest lay in the behaviour of the *phosphate* – not the geranyl residue. During the 1950s there was an increasing interest in the mechanism by which nature synthesises compounds like terpenoids and steroids containing recurring 'isoprene units' in their carbon skeleton, starting from acetate. The discovery of mevalonic acid as an intermediate by Karl Folkers and his group in 1956 gave a big fillip to research in this area, and several of my friends – J. W. Cornforth and G. Popjak in England, Konrad Bloch in the United States and Feodor Lynen in Germany – were deeply involved. I paid no more than passing attention to this field, being absorbed in nucleotide coenzyme studies as well as work on vitamin B_{12} and aphid colouring matters. In August 1958, however, I went on holiday to Lugano with my family and one sunny morning, having just swum in the lake, I was sitting on the hotel terrace with my wife having a Campari when a somewhat decrepit car drew up close by and much to our astonishment the Bloch family emerged from it. Greetings having been exchanged, they joined us on the terrace for a gossip. While thus engaged Konrad said he thought it might interest me to know that there appeared to be a phosphate group in the precursor of the terpenes, which was produced in nature from mevalonic acid. I said to Konrad 'I'm not surprised, but I would bet that the intermediate will be a pyrophosphate' and left the matter at that. It was only later that I remembered Atherton's work, and realised that it really held the key to the problem. Subsequently, of course, the intermediate was identified as *iso*-pentenyl pyrophosphate and this led to the beautiful work on terpenoid and steroid biosynthesis carried out by Bloch, Cornforth and Popjak and by Lynen.

I suppose the main reason for my decision to leave the main brunt of work on polynucleotide synthesis and sequence determination to others was that I was still much involved in coenzyme synthesis, and in the development and refinement of procedures for polyphosphorylation which were a prerequisite for it. Furthermore, I was engaged on structural studies on vitamin B_{12} and in addition to a number of other smaller efforts I was already getting deeply involved in the fascinating problems presented by the remarkable colouring matters present in the haemolymph of aphids. Even although I had for those days a very large research school, there were limitations to what I could tackle!

Following our synthesis of adenosine triphosphate (ATP) our next major triumph in the coenzyme field was the synthesis of flavin-adenine-dinucleotide (FAD) published with G. W. Kenner and S. M. H. Christie in 1952. In the years that followed a variety of other compounds of this type were synthesised including *inter alia* cozymase (nicotinamide-adenine-dinucleotide) and uridine-diphosphate-glucose. It was, indeed, for my work on phosphorylation and nucleotide coenzymes that I received the Nobel Prize for chemistry in 1957.

Ever since my early work on vitamin B_1 in Edinburgh, I retained an interest in the B group of water-soluble vitamins, and did indeed carry out work on some of them in Manchester. One of the most intriguing features of the group was its association with the anaemias. The picture was very confused, and it was not until the 1940s that it became clear that, although such members of the B group as folic acid were involved in nutritional anaemias, the factor involved in pernicious anaemia – the 'external factor' present in liver extracts – was still unisolated. I had not myself taken much interest in pernicious anaemia, and had confined my interest largely to the B vitamins involved in nutritional macrocytic anaemias, but my attention had been drawn to the problem by H. D. Dakin, when I visited him at his home near New York on my way to Pasadena with my wife in 1938. Dakin had been interested in the problem ever since Minot and Murphy in 1926 had shown that whole liver would cure pernicious anaemia, and he had been trying to isolate the material responsible from liver extracts. He had, however, like other workers, made very slow progress indeed since the only way one could test the material was on human patients. Such a test was bound to be inaccurate, but what was (from Dakin's point of view) worse, was that clear cut cases of pernicious anaemia were none too common and clinicians, not unnaturally perhaps, were more interested in curing their patients than in testing Dakin's extracts. When I returned to England I remember discussing the matter with the research group at Glaxo Laboratories Ltd; one of their number, E. Lester Smith, was determined to go ahead on liver extract, and we encouraged him to do so. He slogged on despite every kind of discouragement encountered in the course of testing on human patients and eventually, in 1948, only a very short time after Folkers and his group at the Merck Laboratories in the United States, he did indeed isolate the anti-pernicious anaemia factor Vitamin B_{12}. These two nearly simultaneous isolations of the vitamin were quite independent of one another; but it is remarkable that they should have been so close, when we know that the American group were able to use a microbiological test, while Lester Smith had to go all the way with the much more difficult clinical test procedure. Since I had been associated with the Glaxo isolation work throughout its course, it was perhaps not surprising that I should have been asked if I would undertake a chemical study of it, while Dorothy Hodgkin studied the vitamin by the X-ray method. This I agreed to do, and with my friend and colleague A. W. Johnson we started work. It proved extremely difficult; for one thing we had, during the first year or two of our studies, extremely small amounts of vitamin available to us, and, even more importantly, the molecule proved to be one of almost fantastic complexity. We were able to settle some of its features and learn something about the central part of the molecule by hydrolytic and oxidative studies, but our major contribution lay, perhaps, in the fact that some of our degradation products materially helped Dorothy Hodgkin in her X-ray studies, which in 1955 finally gave the complete vitamin structure. Subsequently, with V. M. Clark, I carried out quite a bit of work on methods which might be applicable to the synthesis of the vitamin, but dropped them, partly because a total synthesis using them would have absorbed a greater part of our research effort than we wished to devote to it, and partly because R. B. Woodward, who had also taken up the synthesis, appeared to me to have a method more likely to succeed (as indeed, with the cooperation of Albert Eschenmoser and his group at Zürich, it ultimately did).

A third major topic of research extending over more than twenty years in Cambridge concerned the colouring matters present in the haemolymph of insects belonging to the family *Aphididae*. My reasons for becoming involved in work on aphid pigments are rather interesting. When I was in Oxford with Robinson I did some work on the colouring matters present in the mycelia of some plant pathogenic fungi of the *Helminthosporium* group. These colouring matters were derivatives of anthraquinone, and, out of curiosity, I went through the literature and listed all the anthraquinones known to occur in nature together with their source and the pattern of substitution in them. It appeared to me that they seemed to fall very roughly into two groups according to their nuclear substitution – those from higher plants on the one hand, and those from fungi on the other. There were, however, two odd ones obtained from insects – carminic acid from cochineal, and kermesic acid from the oak chermes – which seemed to resemble the fungal anthraquinones rather than those from higher plants. This might not seem very remarkable, but I recalled that these insects belonged to the family *Coccididae* whose members are known to contain symbiotic fungi located in special cells called mycetomes. Accordingly, I found myself wondering whether it was the insect or the symbiotic fungi that produced the anthraquinone pigments, and I decided I would look into this when I had some time and opportunity. While in Edinburgh I tried to pursue the matter further. It was, of course, necessary for me to obtain supplies of living cochineal insects since examination of the cochineal of commerce would teach me nothing. I soon found, however, that the authorities were not at all enthusiastic about my importing the insects, and the project went into cold storage until 1939. In the early summer of that year, I drove with my wife and some friends from Manchester to Lake Bala in north Wales for a day's outing and we picnicked by the lake hard by a sizeable stand of foxgloves. As I lay dozing after lunch looking up at the foxgloves, I noticed that one of them had a heavy infestation of black aphids on one of the flower heads; and, I thought, 'aphids are zoologically very close to the coccids – perhaps they too have anthraquinones'. So I took a few aphids and rubbed them between my fingers; sure enough, my fingers were stained. But, oddly, the stain was at first yellowish and then after a short time became red (which was the colour expected of an anthraquinone). So I cut the whole head off the infected foxglove, took it back to Manchester, and had a look at the aphids. It was soon clear that the colouring matter in them was not an anthraquinone, and I confirmed the fact that the coloured substance in the insects did undergo a curious change of colour from a kind of khaki to red within a very short time of its extraction. I also went to the entomological literature and found that the *Aphididae* like the *Coccididae* contained symbiotic fungi. I decided there and then, that I would leave the *Coccididae* alone and look at the *Aphididae* to satisfy my curiosity both as to the nature of their colouring matters and their true origin, i.e. from insect or fungus. But I had to wait until the early summer of 1940 for the next aphid season; by then the war had started in earnest, and I had to put the matter to one side. I resolved to take it up again when the war was over, and so indeed I did.

Among the large group of workers from overseas flooding into my laboratories in Cambridge once the war was over, was a remarkably able Canadian, S. F. MacDonald. He had spent some time with Hans Fischer in Munich, and was a porphyrin-chlorophyll expert doing excellent work in that field. He was a tremendous believer in the use of a hand spectroscope in studying the reactions of coloured compounds, and was immediately fascinated by my account of the aphid colours. A. W. Johnson too was strongly attracted, because I had already got him interested in some other aspects of insect chemistry. So the three of us got to work, and, with a growing band of research students – mainly from overseas – proceeded to unravel the complex puzzle of the aphid pigments. It soon became clear that we had, by so doing, moved into a very large field which absorbed a big effort in manpower for some twenty years. I do not propose to discuss the work in detail. Suffice to say that in general the living dark coloured aphids (red, brown and black) contain a yellowish *protoaphin* converted on the death of the insects by enzyme action to a yellow *xanthoaphin*; this latter is unstable and undergoes conversion in solution successively to an orange *chrysoaphin* and finally to a deep red *erythroaphin*. All of these compounds are complex quinones of a type not hitherto found in nature; they are not anthraquinones. The green aphids – e.g. *Macrosiphium rosae*, the 'green fly' of cultivated roses – do not, contrary to popular assumption, contain any chlorophyll derivatives; they owe their colour to a green quinone related in structure to the aphins. It is of interest that the green fungus (*Peziza aeruginosa*) found on rotting wood also contains a green pigment (xylindein) not very dissimilar in type from that in the green aphids. Much of the later definitive work on the structure of these pigments was carried out in the 1960s, my chief colleague in that phase being D. A. Cameron, now Professor of Organic Chemistry in Melbourne who is still carrying on work in this field (although aphids are not a common pest in Australia, which suffers more from the depredations of coccids!). I never thought when I began work with aphids that it would turn out to be such a monumental undertaking, but it did give a great deal of interest, and my final satisfaction came just about twenty years after I began it, when my young colleague Jonathan Banks established that the colouring matters are indeed produced in the mycetomes of the insects, i.e. they are probably of fungal origin. It would be nice to check the situation in the cochineal insects, and in the lac insects cultivated commercially in India; my guess is that their colouring matters also originate in their fungal symbionts.

The researches on the aphid pigments had their lighter side. If one wanted to get hold of the protoaphins, the insects had to be alive and undamaged; death or injury seemed to liberate an enzyme which at once set in train the series of changes leading to erythroaphins. In the case of *Aphis fabae*, the 'black fly' of cultivated beans (to get them we used to search for a badly infested field, and then pay the farmer to let us cut off the tops of infected plants) we used to place the aphid-infested bean-tops on horizontal shelves in dustbins laid on their side, with a piece of butter muslin stretched across the open end instead of the lid. We then placed a bright light in front of this peculiar piece of equipment; the insects, being phototropic, thereupon withdrew their probosces from the bean plants, walked on to the butter muslin and, hey presto! we had a supply of live undamaged insects. In the case of such insects as the willow aphid *Tuberolachnus salignus*, which lives on twigs and branches rather than on leaves, we had another technique. Aphids, although they can walk quite happily on paper, are completely unable to do so on cellophane; we therefore used to put sloping cellophane sheets under infested twigs then tapped the latter gently. This apparently annoyed the aphids who pulled out their probosces whereupon they fell on to the cellophane and were duly collected. My wife has vivid memories of the early summer of 1948 when we had a heavy infestation of the cherry aphid (*Myzus cerasi*) on a double row of ornamental cherry trees bordering the avenue outside our house. The infestation coincided with the first visit of an Australian cricket side to England after the war. As it happened the gang of research students whom I sent out to climb and 'delouse' the trees were Australians; I understand that they used to descend from their perch in the trees every half-hour or so, to have some tea and keep up with the Test match on the radio! For the aphid work we needed very large amounts of insects, and so used to collect them by the kilogram. In a poor aphid year we sometimes had a good deal of difficulty in getting particular species. I recall that in one year we put an advertisement in the local newspaper asking people to let us know if they came across aphid infestations, so that we could collect the insects. We only did it once; no-one gave us any information, but we had a flood of envelopes from helpful people enclosing a leaf with two or three aphids on it, and about twice as many crawling all over the outside of the envelope. I don't think the postal authorities thought much of the result either.

6

The post-war years. Involvement in science policy and international activities

—

As far as we in Cambridge were concerned, the war effectively ended with the collapse of Germany in May 1945. True, the struggle with Japan continued, but it was rather remote from us. Although air-raids from Germany had ceased, we were, until the end of the war in Europe, subject to the continued threat of flying bombs and the more dangerous rocket-propelled V-2s. With that menace removed, we seemed effectively at peace, although all or most of our wartime restrictions continued and life was certainly austere. In the autumn of 1945 I received an invitation from the Swiss Chemical Society to visit Switzerland in early December, and lecture in Zürich and Basle on my researches on *Cannabis*. This was my first trip abroad since the outbreak of war, and I was very excited about it. I reported at Northolt airport on a bitterly cold morning to fly in a Swissair DC-3 to Zürich. Fog delayed our departure for an hour or two, much to the pilot's concern – he was clearly afraid that if he didn't get to Zürich before dark he might not be able to land there. However, we finally took off, and when we arrived over Zürich it was already dark but clear and frosty. Looking down from the plane on a city blazing with lights and flashing advertisements, I could hardly believe it – so accustomed had we become to the gloom of the blackout in Britain. Once landed, the shock was even greater. I find it hard to express my feelings at seeing again well-lit streets and shops, with glittering snow all around – there were sweet stalls, and you could even buy bananas in the greengrocers' shops! I had almost forgotten what the world had been like before 1939! But it was a magnificent feeling – this was peace, and soon all Europe would be like this again (although, in fact, it was to take longer than I expected). It was a most successful visit; my lectures went well, and I was able to renew my contacts with Professors Reichstein, Ruzicka and Karrer. I also, of course, made touch with my old research colleagues from London and Manchester – Marguerite Steiger now running her family's pharmaceutical business in Zürich, and Hans Waldmann with Hoffmann La Roche in Basle. I also renewed acquaintance with two old Oxford friends – Emil Schlittler and Rudolf Morf; the latter I was to see much of in later years through our joint concern with the International Union of Pure and Applied Chemistry, of which he became Secretary General in 1955 in succession to Professor Delaby of Paris.

In February 1946, about two months after my return from Switzerland, I was off to the Continent again – this time to Germany. My old friend Bertie Blount, whose family had always been army people, was unable to resist the desire to get into uniform during the war and had a distinguished career in the army. At the end of the war he had the rank of staff colonel in the British Zone of Occupation, where he was responsible for the control of chemical and biological research. Although theoretically located elsewhere, he soon realised that Göttingen was the scientific centre of the Zone and spent most of his time working from the Aerodynamische Versuchsanstalt (whose occupying staff included a resourceful airman named Ronald Purchase). One of his great interests was to collect some of Germany's leading scientists together there, and, in due course, to re-create the old Kaiser Wilhelm Gesellschaft in a new form as the Max Planck Gesellschaft. This Göttingen group became part of the Control Commission's responsibility, and Bertie remained with it for a few years before leaving the army and returning to civilian life. One of his great interests was to see German science – which we both knew well from our student days – started up again in the devastated Germany of those days. So it was that, in February 1946, I was asked by the Foreign Office (at his instigation) to visit the British Zone of Occupation and see what might be done. At that time there was no way in which a civilian could do such a thing, so I was given a status equivalent to brigadier, dressed up in an ill-fitting army battle-dress, and put on a rather decrepit old Dakota which was ferrying supplies and occasional passengers to Buckeburg in north-west Germany. There we landed on what looked like a sea of mud, and I was met by Bertie; we drove to Göttingen where I was put up in the officers' mess. The mess was rather spartan, but reasonably comfortable, and we were well looked after. Certainly, the contrast between it and what I saw of the living conditions of the German civilians in Göttingen and in other parts of the country I visited was striking. Germany was in a shocking state; there had been appalling destruction in the cities (although not in Göttingen nor in many other small towns), food everywhere was scarce and bad, money had become more or less valueless, and daily life was conducted under a kind of barter economy in which the common currency seemed to be cigarettes.

Using Göttingen as our base, we made visits to a number of main centres of academic activity in the British Zone – Cologne, Hamburg, Kiel – and I renewed acquaintance with a goodly number of chemists, many of whom I had known before the war. These included Brockmann, Windaus, Bayer, Alder and Diels among others. The tour was not without its amusing features. I still recall the look of astonishment and disbelief on the face of Otto Diels when he discovered that the 'British general' who, he had been told, wished to interview him turned out to be me, and the not dissimilar reaction of Kikuth at the Elberfeld research laboratories of the old I.G. Farbenindustrie who, on asking me rather resignedly on my arrival in uniform what I wanted from them, was told that I really didn't want anything but a cup of tea and a chat.

Apart from a fleeting visit paid by Bertie Blount during the last days of the war, when he was attached to S.H.A.E.F. (Supreme Headquarters Allied Expeditionary Forces) and on his way eastwards to Leipzig, neither of us had had any contact with Walther Borsche since before the war. We knew that he had been dismissed from his post, but, although he was living in the country at Friedberg (Hessen) when the war ended, we had reason to believe that he and his wife were back in Frankfurt (although not in the university) by February 1946, occupying rooms in the house of a scientific friend (Dr Rajewski). Borsche fell foul of the Nazis and lost his academic position although he had no trace of Jewish blood in his ancestry; but, like his old friend Adolf Windaus, he made no attempt to conceal his hatred of everything the Nazis stood for. Bertie and I decided we would begin our scientific tour by going down to Frankfurt to see our old teacher. Frankfurt a.M. was, of course, in the American Zone, but we managed to think up some excuse for going there, obtained the necessary permits, and set off by road.

In Frankfurt we had first to present ourselves to American headquarters in the Palmengarten. My chief recollection of our call there is the remarkable behaviour of the (Puerto Rican) sentry at the main gate, who insisted on examining our papers page by page while holding them upside down.

We located the Borsche family without difficulty, and were able to visit and to spend a happy hour or two with our old professor and his wife. They were in very good shape, and seemed to be as reasonably comfortable as one could be in a German city at that time. Frankfurt was in a shocking mess. To me, particularly depressing was the destruction of the Altstadt. The cathedral was still more or less intact, but I remember standing on the steps of what had been the Römerhaus – only the facade was still standing – and looking across a wilderness of rubble towards the Main. Everything I had known so well in the old days had gone – the Roseneck, the Fünf-Fingergasse – everything! One morning I thought I would like to visit the university at Bockenheim, and have a look, not only at the laboratories, but also where I had lived; I even thought I might be able to trace my old landlady in the Königstrasse. We were lodged in a hotel in the Bahnhofsplatz about a mile or so from the university, so I thought I would walk. I certainly got a shock. On the way to the university I don't suppose I saw more than a dozen undamaged houses. In Bockenheim itself Königstrasse, in which I had lived, had simply vanished and there was also a good deal of damage to the university buildings. Part of the chemical laboratory building where I had done my research was still standing in the Robert Mayer Strasse, and, as I could see on approaching, it was being used, since students were passing in and out. Although I didn't expect to find anyone there whom I knew, I decided to go in and have a look around. This I did, and, although my uniform attracted some attention, no-one interfered and I was able to look at my old laboratory – still just as it used to be – and wander through the building. In the course of this tour I went down to the basement, where the store which issued chemicals and equipment used to be. It was still there and still functioning as in the old days, with a queue of students filing past to purchase or borrow things. What really surprised me was that I could see that the white-coated figure behind the store counter was the same Herr Möller who had been storekeeper in my day. So I quietly joined the queue of customers, and, in due course, came up to the counter. Möller looked up, gazed at me in silence for nearly a minute, and then in his ripe Frankfurt dialect said 'Good God, who would have believed it – it's Herr Todd.' And with that he pulled down the shutter over the counter, emerged from the side-door, seized me by both hands and said 'Come in, come in – this calls for a drink!' So in I went and sat on one of the two stools in the store. Möller, meanwhile, took two beakers from a cupboard, put a generous amount of laboratory alcohol into each, diluted them with a roughly equal quantity of distilled water, handed one to me, then sat down on the other stool. We toasted one another and the old days several times in this ghastly potion, and then Möller began to chat about the laboratories and their inhabitants. 'Herr Todd', he said, 'in our days we had chemists in this place, eh? You remember them – von Braun, Borsche and the others! Ah! Things have changed! Do you know, some of the people the Nazis sent here were scientifically so small that you could hardly see them!'

While in the American Zone we drove down to Darmstadt (although strictly we had no authority to do so), in order to see another chemical friend, Clemens Schöpf, who was professor at the *Technische Hochschule* there. He and his wife were well, but living in a rather battered house standing in a sea of rubble. Thereafter we returned to Frankfurt, and set off northwards to Düsseldorf to begin our tour of some of the places in the British Zone where, as in Göttingen, there were signs of a renascence of scientific activity. On the way to Düsseldorf we chose a route which took us through old haunts in the Taunus hills, and, in particular, to the little town of Idstein, where I knew that the family of my former co-worker, Anni Jacob, lived, and where her brother-in-law was the village pharmacist. We called on him in the forenoon of a crisp sunny day, and were greeted with great enthusiasm. The entire Jacob family was summoned, and Anni's brother-in-law and nephew went off to the bottom of the garden and dug up several bottles of their best wine, which they had carefully buried to protect it from the American occupying troops. We had a hilarious time, and, much later in the day, departed in a somewhat tattered condition, complete with some wine and a bottle of home-made spirit rather flatteringly called 'gin', the production of which seemed to be a flourishing cottage industry in Idstein. (That bottle ultimately fetched up in the mess at Göttingen, where it was labelled 'Echtes Toddka' and stood there in a cupboard for about two years before it (allegedly) exploded.) We arrived late at night in Düsseldorf, but my recollections of the journey and our arrival are a bit hazy! After discharging some official duties with the British authorities in Düsseldorf, our route took us on to Hamburg, Kiel, Cologne, Bonn and Elberfeld in succession. Thereafter we went back to Göttingen and I returned to England.

During this trip I saw enough the realise that, with the kind of encouragement that Bertie Blount was giving to the German scientists, it would be possible to get science in the British Zone into reasonable shape quite quickly once the general economic situation improved. I also learned, to my great relief, that the two great German chemical encyclopaedias, Gmelin's *Handbuch der Anorganischen Chemie* and Beilstein's *Handbuch der Organischen Chemie*, had been rescued from Berlin by the efforts of Blount and Professor Roger Adams who was, as a civilian, performing the same kind of duty in the American Zone of Occupation. The editors, staff, and most of the material belonging to these works were now located in the West – Dr Pietsch and Gmelin in Clausthal-Zellerfeld, and Dr Richter and Beilstein in Höchst, near Frankfurt. They still had many problems to face, not least among them being the procurement of current scientific literature, which they could not afford to buy abroad. It seemed to me that I would probably be back in Germany again before long, and in this I was quite right.

Within the next eighteen months I paid two further visits, based in each case on Göttingen as before. On the first of these trips I was accompanied by H. W. Thompson (now Sir Harold Thompson) and H. J. Emeleus. Thompson was already an old friend of mine, and an even older one of Bertie Blount; we had all three studied in Germany at the same time; Emeleus, of course, was now my colleague in Cambridge, and he had studied in Karlsruhe at about the same time as I had been in Frankfurt. We three, like Bertie Blount, had the great advantage of being fluent German speakers, which made everything go more smoothly. During this visit we had a gathering in Göttingen of all the German chemists who could be mustered, at which the new Gesellschaft Deutscher Chemiker was formally set up, replacing the virtually defunct Deutsche Chemische Gesellschaft. The meeting, which included participants from both the French and American Zones was, I think, the first scientific meeting held in Germany after the war. It was a great success, and we had a symposium at which Thompson, Emeleus and I read papers on what had been happening in chemistry outside Germany. Emeleus and I had a good look at the problems of Gmelin and Beilstein, and I then created, unilaterally and without authorisation, a Beilstein–Gmelin Commission of the International Union of Pure and Applied Chemistry (subsequently whitewashed by Professor Delaby, the Secretary General of the Union) and wrote on its behalf to the publishers of all the main chemical journals in the world, asking them to donate copies of their publications to each of the two Institutes. The response was magnificent; almost without exception, they sent two copies of each issue to me in

Cambridge and I had them transmitted to Germany through official channels. It involved Emeleus and me in a lot of work until mails were reliable enough for journals to be sent direct from the publishers, but I like to think it was worth it; had it not been done these two great encyclopaedias might well have foundered and disappeared. Of course, all was not smooth sailing – we had to cope with disputes between Dr Pietsch and his German governing body about what weight was to be given to the history of chemistry, and there were numerous alarms and excursions over attempts by publishers in England to get hold of Beilstein. But it was all good fun and very interesting, as also were the visits we made to German industrial research organisations on our 1947 visit. Looking back on things now, it seems to me that Britain really did a very good job for science in its zone of occupation, and I believe that much of the credit for this must go to Bertie Blount; his services in this respect are remembered with gratitude in Germany, but have received less than their due recognition at home.

Prior to 1946 I had little contact with national or international scientific organisations, and my interest in chemical education did not extend much beyond the day-to-day involvement with teaching in my own department. However, the visits to post-war Germany, which I have just mentioned, introduced me to some of these matters and in the years that followed I became increasingly involved in them and in the promotion of chemical research by the award of postgraduate scholarships and fellowships, serving on a variety of charitable organisations concerned with such matters. Of these, one of the more interesting was the Salters' Institute of Industrial Chemistry which was a major activity of the Worshipful Company of Salters, one of the twelve great livery companies of the City of London. The great livery companies of London are the descendants of mediaeval craft guilds. Although many of them – like the Salters – have long since lost their significance, they own much city property and spend large sums charitably, mainly on schools, almshouses and the like. Early in this century the Court of the Salters' Company decided that it would like to devote at least some of the Company's wealth to promoting the chemical industry, as being perhaps the nearest equivalent to its ancient, but vanished, involvement with salt production and distribution. Accordingly, it established a body known as the Salters' Institute of Industrial Chemistry, to provide assistance by way of grants to boys wishing to study chemistry, and by making awards at the postgraduate level to young men wishing to pursue research preparatory to taking up an industrial career. Admission to the livery of the Salters (which numbered only about 120 in all) was almost exclusively by patrimony, and so it happened that when, in 1946, the Company sought to reactivate the Institute (which had been inactive during the war years), they found that the livery contained no members concerned in any direct way with chemical education or research who might assist in its work. The Court of the Company then decided to break with its traditions, and to invite two chemists – myself and Charles Goodeve – to join the livery and help run the Institute. This we did, and, although because of war damage to its London properties the Company for a time had only very limited funds available, the Institute restarted its operations and has grown steadily in its scope ever since. My association with the Salters' Company has always been a happy one; I served as Master of the Company in 1961–62 and my son and son-in-law continue the family connection as liverymen and members of the Institute Committee.

Both in Britain and in the United States science was fully harnessed to the war effort between 1939 and 1945 with the spectacular results we all know. In Britain guidance for the overall effort was provided by the Scientific Advisory Committee to the War Cabinet. This committee, which was chaired by the Secretary to the Cabinet, Sir Maurice (later Lord) Hankey, was quite small consisting of the President of the Royal Society (from 1940–45 my father-in-law, Sir Henry Dale), two of the Society's Honorary Secretaries and the heads of the Medical and Agricultural Research Councils and the Department of Scientific and Industrial Research. The impressive contributions made by science during the war years led to a feeling in government, as well as in scientific circles, that perhaps a somewhat similarly organised scientific effort would be equally valuable in peace, and especially during the period of reconstruction and re-orientation of effort which would follow immediately after the end of hostilities. Shortly after the end of the war, a committee under Sir Alan Barlow (the Barlow Committee) was set up to look at some aspects of this, and especially at our needs for scientific manpower. Amongst other things, it recommended the setting up of two advisory bodies at the highest level – an Advisory Council on Scientific Policy dealing with civil science and technology, including manpower requirements, and a Defence Research Policy Committee to deal with defence matters. It was envisaged that these two bodies would have a common chairman, so that proper liaison would be ensured, and it was suggested that the chairman would be employed as a full-time civil servant. These recommendations were accepted, and, in 1947, the Advisory Council on Scientific Policy was set up 'to advise the Lord President of the Council in the exercise of his responsibility for the formulation and execution of Government scientific policy'. Sir Henry Tizard was appointed to be simultaneously Chairman of the Advisory Council and of the Defence Research Policy Committee. Much to my surprise I received from the Lord President of the Council (Herbert Morrison) an invitation to join the Council as one of the original members. The invitation surprised me, partly because as a chemist I had not been heavily involved in the war effort – certainly less so than many physicists – and partly because I did not consider myself at all knowledgeable on policy matters; but the prospect was attractive and I accepted the invitation.

I suppose it was my membership of A.C.S.P. that first stimulated my interest in the interaction between science and government which is commonly described by the rather loose expression 'science policy'. Until I joined it, I had taken little interest in affairs outside chemistry and the universities with which I had been associated. Certain it is that from about 1947 onwards I became increasingly involved in the problem of providing adequate scientific and technological advice to government, and this has remained a major interest in my career since then, both on a national and international level. The need for governments to be given all available scientific and technological information on any given topic, together with the scientific implications of that information (insofar as they can be foreseen), if they are to take wise decisions in the field of public policy is now, I think, well recognised. The problem is to devise the proper machinery for doing so. That machinery must vary from country to country according to the system of government, but even a generation after recognition of the need, no wholly satisfactory answer has yet been found for any country.

As originally constituted, the Advisory Council on Scientific Policy (A.C.S.P.) had twelve members in addition to its chairman. Of these, three were academic scientists, two were scientific industrialists, and one was an officer of the Royal Society nominated by the President. In addition, there were the three secretaries of the Research Councils, the Chairman of the University Grants Committee, a senior Treasury official, and a chief scientist from a government department. It was later enlarged to include a scientist representing the interests of atomic energy and the Director of the newly formed Nature Conservancy. All appointments were personal, i.e. no alternates were allowed and members were expected to give their own views, and not simply peddle briefs on the part of any organisation to which they belonged. It will be observed that official and non-official members were finely balanced so as to avoid officialdom ruling the roost. The secretariat, which was quite small (in my view, too small), was attached to the office of the Lord President of the Council, the minister responsible for the Research Councils and for general advice to the Cabinet on scientific and technological policy. In the original A.C.S.P., Solly Zuckerman and I were very much junior to the other members with the exception of E. W. Playfair, the Treasury representative, and we tended to play the part of the opposition from time to time, and this was probably no bad thing. I continued as a member and served on various panels or sub-committees until I retired by rotation in 1951 and, during that time, I learned a lot from Henry Tizard about running committees, about balancing opposing viewpoints, and reaching proper decisions. Tizard was an immensely able, and withal an extremely friendly, man with an impish sense of humour and a fine wit. Many are the stories told about him; one of his practices which I remember was his method of choosing members for any A.C.S.P. sub-committee which appeared to have a tiresome and unattractive remit. Tizard used to begin the operation by looking around the conference table and saying 'Well now! Who isn't here today?'

My resignation in April 1951 gave me little respite from A.C.S.P. On 31 March 1952 Tizard retired. Only a few months before he was due to leave, a Conservative government under Churchill came into power again and Lord Woolton, who had made a great reputation as Minister of Food during the war, became Lord President of the Council. I knew and liked Lord Woolton (known, to many of his younger friends at least, as Uncle Fred) who was a fellow member of the Court of the Worshipful Company of Salters. I had a note from him one day in February 1952 asking me to come along and see him at the House of Lords. When I got there, he said there was no point in beating about the bush – would I accept the chairmanship of A.C.S.P. in succession to Tizard? He explained that it had been decided that the new chairman should be an independent scientist working part-time, and that the office should be separated from the chairmanship of the D.P.R.C.; liaison between the two bodies would be arranged by having each chairman (or his nominated representative) sit on the other body. This arrangement was, of course, more acceptable to me than that under which Tizard had operated, for I was quite determined not to join the civil service. However, just at that time my researches were forging ahead – we had just solved the nucleic acid structure, completed the synthesis of flavin-adenine-dinucleotide, the first of the complex nucleotide co-enzymes, and all sails were set. So I told Woolton that I really was very busy, and that I was trying to cut down commitments rather than take on any more. We talked around the subject for a bit, and then he said 'I'm asking you to do something important for the country. Do I take it you don't want to help your country? After all, you can have my authority to resign from any other government commitment you may have, and stick simply to A.C.S.P. How about it?' Well, I didn't have much option after that, so I said 'All right, I'll do it – but although I will accept an allowance to cover my out of pocket expenses, I will not accept any salary or honorarium, since I feel that, in the position you wish me to occupy, I should be wholly independent of the government or civil service.' He said 'Done!' and that was that. I became chairman of A.C.S.P. and was able to resign the chairmanship of the Chemical Board and membership of the Advisory Council to the Ministry of Supply, and get rid of one or two other small but at times irksome commitments. I never regretted taking on A.C.S.P., whose chairman I remained until just before its dissolution on the advent of a Labour administration in 1964. I shall have occasion to mention the events leading to my resignation later, but I believe that, if we had had associated with A.C.S.P. a full-time scientific adviser in the Cabinet Office, it would have given us a more effective instrument for the development of science policy than anything we have had since. Meanwhile, all sorts of other things had been happening.

In the summer of 1947 I had a visit from an old friend, Joe Koepfli, whom I had not seen since our stay in Pasadena in 1938. He was in London for a spell as scientific counsellor at the U.S. Embassy, and he introduced me to Earl Evans, one of the two professors of biochemistry at the University of Chicago who was going to follow Koepfli in the London position during sabbatical leave from Chicago in 1948–9. The possibility that I might spend a term as a visiting professor at Chicago during his absence was raised by Evans, and, in due course, I was invited to spend the autumn term 1948 as professor of biochemistry in Chicago. The idea of visiting the United States again and renewing contacts was very attractive, so I accepted. I hoped that Alison would be able to accompany me, but having three young children who couldn't come with us was a bit of a complication. However, we managed to find a suitable temporary housekeeper to look after things in Cambridge for part of the term at least, so that I was able to go on ahead, and

Alison joined me in Chicago at about the end of October. I had recently become chairman of the Chemical Board of the Scientific Advisory Council of the Ministry of Supply, and I felt it would be a good thing to visit the corresponding American organisation with which we had become closely associated during the war. Accordingly, I travelled to Washington in mid-September 1948 with Dr Fred Wilkins (then Controller of Chemical Defence at the Ministry of Supply and later to follow Sir Harry Jephcott as chairman of Glaxo Laboratories Ltd), and attended various meetings with our American colleagues there and in Baltimore. I regret to say that my only vivid recollection of these meetings is one of a dinner in Baltimore given by senior military and naval staff. As is not uncommon at such events, the bourbon flowed freely before we got as far as the dinner table, where, to my astonishment, I watched a high-ranking officer sitting opposite me solemnly add sugar and cream to his consommé. This remarkable mixture he consumed during the remainder of his meal, giving no indication that it appeared to him to be in any way different from the coffee which his colleagues were drinking!

At the end of the month I proceeded in gentlemanly fashion from Washington to Chicago by the Baltimore and Ohio railroad; it really was quite a delight to travel at a leisurely pace up through the Cumberland Gap and on to Chicago. There I was met by Crawford Failey, the second professor of biochemistry who was to be my colleague for a term, and we drove off to the University Graduate Club on 59th Street, where I was to live during my stay. I remember that drive well. For one thing we went south on the lakeside driveway at a furious speed three abreast and driving almost bumper to bumper, in a (to me, at least!) most alarming way. Secondly, the first instruction given me by Crawford as we drove along was that I should avoid walking on the Midway (an open space with grass and trees close to the university buildings) after dark, and that, if I did have to do so, I should carry ten dollars with me. He told me that this was important, since I would probably be held up; if I had less than ten dollars I would be beaten up, while if I had more it would be a sheer waste of money! So it seemed there was something in the legends of Chicago after all. To be fair, however, I must say that I never encountered any unpleasantness while I stayed in Chicago, even although there were areas near the university with very dubious reputations, and a number of nasty incidents were reported there during my stay. What did rather surprise me was the indifference displayed by most people I met to violent crime; it seemed to be accepted as part of everyday life.

I liked Chicago. It was an exciting, lively city and Alison and I made many friends – Crawford and Christine Failey, and Phillip Miller, professor of medicine and his wife Florence, together with my organic chemical counterpart, Morris Kharasch and his wife Ethel May – all of whom were particularly kind – and we saw a lot of them. In the month or so before Alison arrived I must have been very busy for, apart from my two lectures each week in biochemistry, I managed to fit in a couple of trips to New York, a visit to the University of Illinois, and a week at Notre Dame University giving the Nieuwland Lectures. When in Chicago, there was always billiards after lunch at the Graduate Club with Morris Kharasch and Warren Johnson. On Wednesday evenings Morris used to have a seminar, and it always concluded with a challenge match at bridge – Morris and myself versus any two graduate students who would take us on. Morris used to say that he and I made a perfect bridge partnership. As a theorist accustomed to making ingenious theories with very little evidence, he felt he should control the bidding; I, on the other hand, being an expert in doing successful experiments with very little material, was the ideal person to play the hands. Whether that view was valid or not, he and I usually won, consuming a good deal of his favourite tipple, Southern Comfort, while we were doing so. The biochemistry department was a good place to work in, and in addition to Crawford Failey, who was much tied up in administration, we had Frank Westheimer and Konrad Bloch, both of whom later went to Harvard, and who have been fast friends of mine ever since the Chicago days. I also got to know Robert Mulliken quite well, although his theoretical interests were rather far removed from my own. The University of Chicago was a lively place in those days, and I liked Hutchins who was then President. Particularly during my bachelor existence in the first half of my stay, I got around quite a bit in the university and in the city of Chicago. I particularly recall how on Sundays (the Graduate Club having no breakfast or luncheon facilities) I used to proceed in leisurely fashion downtown to the Tavern Club, of which I was a member (thanks to Crawford Failey), and have a vast brunch high up in a skyscraper with a breathtaking view over the lake. It was during my Chicago period that the almost legendary presidential election between Truman and Dewey was fought. I actually got hold of the famous *Chicago Tribune* edition which jumped the gun by prematurely announcing Dewey's victory; that edition was actually on sale for a brief period in the early morning following the election. In the run up to that election, which somehow seemed to be simultaneous with a lot of other elections in Chicago, I was astonished by the libellous statements about individual candidates which blazed forth from posters put up by their opponents; evidently American law is much more permissive than English.

Shortly before the end of our visit Alison and I flew out to Los Angeles and spent a day or two visiting old friends like the Paulings, Koepflis and Niemanns in Pasadena. It was a very jolly reunion but I am afraid that the massive growth of CalTech and Pasadena made them, although still attractive, less so than they were in in 1938. Los Angeles, then getting afflicted by smog, had deteriorated a good deal more, and, to me at least, had no longer any attraction at all. We returned to Chicago feeling that perhaps our decision in 1938 was the right one, and shortly thereafter we left for the east for a few days in New York and the Boston area. We stayed in Cambridge, Massachusetts with Oliver and Alice Cope, close friends of Alison's family, and had a marvellous few days meeting a lot of friends both of Alison and myself. Oliver Cope, a surgeon at the Massachusetts General Hospital, had worked with Alison's father; indeed the whole area seemed to be full of former associates or pupils of Sir Henry Dale, all of whom knew Alison. On the chemical side I was able to meet again R. B. Woodward; this, of course, I greatly enjoyed.

Although Bob Woodward was ten years younger than I and

we lived in different continents, he and I were intimate friends for some thirty years until his sudden and most untimely death in the summer of 1979. He was one of those very rare people who possessed that elusive quality of genius, and was certainly the greatest organic chemist of his generation, and possibly of this century. I believe I first met him fleetingly at a party in Cambridge Massachusetts when Alison and I stopped there on our way home from Pasadena, but my attention was first drawn to him by a paper he published in 1941 (at the age of twenty-four) on the ultraviolet absorption spectra of $\alpha:\beta$-unsaturated ketones; it seemed to me to herald a breakthrough in the use of spectroscopy in the study of molecular structure. In this I was right, and Woodward went on from strength to strength. During the war we were in contact and were both involved in the Anglo-American penicillin investigations, but we did not actually meet again until 1947 when he made a brief visit to England. So my visit to Harvard in 1948 was the more memorable because it really marked the beginning of my close association with Bob Woodward.

Back in Cambridge, things were very much on the move. Thanks to Ralph Gilson's efforts, the University Chemical Laboratory had undergone a remarkable transformation; all the anachronisms like gas lighting had gone, laboratories had been remodelled and re-equipped, library, stores and other services had been enlarged and completely reorganised, and a totally new spirit of confidence and enthusiasm pervaded the school. The flood of overseas and of British doctoral and postdoctoral students seeking to join us kept growing to the point of embarrassment and we had indeed to commandeer some of our teaching accommodation to cope with it. By 1948 we had – in addition to British – Australian, New Zealand, Polish, German, Spanish, Indian, Canadian, South African, American, Chinese and Singhalese students, all forming a closely knit group and all working productively. Socially the laboratory was a good outfit too. We used to have a joint cricket team with the technical staff which played for some years in the Cambridgeshire Cricket League, and, in this and other social and sporting activities, there was no distinction made between the most junior laboratory assistant and the professor. My work was increasingly attracting the attention of biochemists, especially in America, where biochemistry teaching was much more grounded in chemistry than it was (at the time) in Britain; indeed, at the time of the International Congress of Biochemistry held in Cambridge in 1949 I remember being approached by a number of American participants, including Carl and Gerty Cori, to persuade our biochemists to get closer to American practice, which seemed to produce much better research schools! There was quite a ferment in Britain at that time about the relationship which should exist between chemistry and biochemistry, and I was viewed with quite unjustified suspicion by leading British biochemists, because I was unable to regard organic chemistry and biochemistry as totally distinct disciplines with fixed boundaries. It was about this time, incidentally, that a young American biochemist came to my laboratory as a postdoctoral fellow and, on introducing himself to me, said 'Professor, I have come to you to be saved.' Some thirty years have passed since then, and it has given me, at least, the greatest pleasure to see the change of heart from those days – a change which has brought chemistry and biochemistry much closer to one another to their mutual benefit.

While in Chicago, and indeed for some time before that, I had been subject to rather tiresome bouts of indigestion and early in 1949 matters came to a head rather suddenly, and I had to undergo major surgery for removal of my gall-bladder which was completely blocked by a large lump of beautifully crystalline cholesterol. The Dyestuffs Division Research Panel of Imperial Chemical Industries Ltd used to meet on the first Friday in each month at Blackley Works, Manchester, and its three external members – Robert Robinson, Ian Heilbron and myself – used to travel from London by rail on the evening before and stop over at the Midland Hotel in Manchester. When I had recovered from the gall-bladder operation sufficiently to go up again to a Panel meeting, I found myself travelling up on the evening train with Robert Robinson. Robinson, my former teacher and now a close friend, had, as one of his more endearing characteristics, a habit of reacting emotionally and almost violently at times to comments on chemical matters which were brought suddenly to his notice. At the time of which I am writing, he was in the throes of his large-scale effort to synthesise cholesterol – a rather fashionable pursuit which then occupied a number of competing groups throughout the world. Robert proceeded to recount to me that evening a particularly nasty snag that was blocking his path. After we had discussed it for a bit I said 'I'm sorry you are having difficulties – I don't suppose you know it, but I have just completed a synthesis and isolated cholesterol absolutely correct in structure and stereochemical configuration.' He rose to the bait like a trout to the mayfly, and almost shouted 'What do you mean? Why was I not informed that you were working on cholesterol? How long has this been going on?' I replied 'Don't worry, Robert, it's just a little something I did entirely on my own in my spare time' – and with that produced my beautiful gallstone, which I was carrying in my waistcoat pocket. Robert's wrath vanished as quickly as it had come, and was replaced with a roar of laughter; when we got to Blackley next day, I heard him recounting my successful synthesis to all and sundry.

In the summer of 1949 I received my first honorary degree – Doctor of Laws – from the University of Glasgow. I was deeply touched, for not only was it my first academic honour, but it was bestowed by my own university where I began my career. Just a year later I received a second – the honorary degree of Dr. rer. nat. from the University of Kiel. This I also esteemed greatly, since it was bestowed, I suspect, partly in recognition of what I had done for the German universities, and especially for German chemistry, in the dark days just after the war. The spring of 1950 was also the occasion of a hilarious visit to Spain, with Harry Emeleus, to celebrate the 10th Anniversary of the Consejo Superior de Investigaciones Cientificas (of which body we were both Consejeros de Honor). The celebrations were on a lavish scale, but two particular functions were especially memorable. The first was the commemorative act in Madrid where the handsomely bound volumes containing the work of each of the numerous institutes of the Consejo were presented to General Franco. This was held in a large hall fitted up with floodlights trained on a platform on which was a long table with a reading desk in the centre and a semicircle of chairs behind it. We were all assembled in good time, and were kept amused by the efforts of the electricians who were clearly having trouble with their lighting arrangements. About half an hour after the advertised time for the ceremony the Caudillo and his cabinet – all in magnificent white uniforms – arrived in a fleet of cars, flanked by an escort of about a hundred soldiers or police on motor-cycles. The bands struck up, and Franco and his ministers processed into the hall. Unfortunately, just as the floodlights were turned on the main fuse blew, and for a minute or two chaos supervened. When they were turned on again, all the white uniforms had got on to the platform, but the effect was rather spoiled by the fact that some of the ministers seemed to be waving to friends in the audience, the main spotlight was slightly off-centre, and the platform was swarming with photographers all pushing and jostling to get a good picture of the Caudillo. It was all rather like a comic opera version of a Nazi rally. Order was finally restored, the Caudillo took up his stance at the reading desk, and, one by one, the institute directors came forward, each with one or more attendants bearing stacks of books which were deposited on the table in front of the General. Now, Franco was not very tall and the pile of books grew rapidly – so much so that by the time we were about half way through the ceremony, Franco could no longer be seen from the part of the hall in which I was seated; it really was a hilarious meeting.

The second highlight of the Consejo celebrations was the excursion to Segovia. I do not think General Franco himself was there – at least I do not recall seeing him – but quite a substantial number of his ministers took part. We were all loaded into large black cars in Madrid, and drove in procession to Segovia. All along the route, at intervals of about a kilometre, there were pairs of civil guards who presented arms. When we got to Segovia we were received by the provincial governor and the mayor, and drove up to the Alcazar which had been specially beflagged for the occasion. On alighting, we entered the great courtyard, each being presented on the way in with an alforja (a kind of brightly coloured saddle bag) containing in one pocket a sheep's milk cheese, and in the other some bread, a knife, and a glazed earthenware jug of about a pint capacity. The reason for this was soon apparent. Along two sides of the courtyard were rows of sucking pigs roasting gently on spits, and on the third were skins of wine suspended over large horse or cattle troughs. The centre of the courtyard was occupied by troupes of regional dancers from different parts of the country. The wine skins were slashed, the dancers danced, and we all got busy on the wine and bread and cheese. I suppose this went on for about half an hour, by which time all inhibitions had been removed, and we all tottered (literally) into the great hall where lunch – vast quantities of barbecued sucking pig and still more wine, was consumed. Thereafter, we got back into our cars, and slept peacefully until we arrived at the Escorial where, believe it or not, we were met with copious draughts of brandy. My recollections of the rest of the trip back to Madrid are a bit hazy.

The incidents I have described were amusing, but in no way detracted from a most impressive celebration of the Consejo Superior's birthday. In my experience, the Spaniards are extremely friendly and hospitable people, and, when they decide to have a celebration, it is always on an impressive scale. A further example occurred in 1953 when my wife and I attended the Jubilee (Bodas de Oro) of the Spanish Chemical Society in Madrid, and again later, when I received an honorary doctorate from the University of Madrid in 1959.

I spent a good deal of time overseas in 1950 for, apart from the short trips to Spain and to Germany which I have mentioned, it was in that year that I made my first visit to Australia. I received an invitation from the Royal Australian Chemical Institute to lecture in the state capitals, and also one to be Visiting Professor of Chemistry at the University of Sydney for the Australian spring term (September–December 1950). I had always wanted to visit Australia, and, although, unfortunately, Alison was unable to accompany me, I accepted the invitation and set out by air in mid-August. The flying-boat service had ceased by then, but I had a marvellous trip out on a Qantas Constellation. From London we flew to Rome, where we were provided with picnic lunches, put on a bus, and given a tour of the city before re-embarking for Cairo, where we arrived late in the evening. We had late dinner, and it was such a beautiful moonlit night that, instead of going to bed, I, with one or two fellow passengers, hired a car and visited the pyramids, where we spent a marvellous couple of hours. At dawn we took off again, and with a couple of intermediate stops came down very late at Singapore where we were Qantas guests until a day and a night later we took off again early in the morning for Darwin. The weather was fine, and the volcanic peaks of Java, the tips of which seemed almost level with the plane, were quite magnificent. Shortly after leaving Singapore the captain announced that, after much haggling, an agreement about landing rights had been signed by Australia and Indonesia the previous day, and that he had been asked to be the guinea-pig and call at Jakarta. This we did; I would hardly describe our welcome as warm, and we were all locked up in a kind of shed with an armed guard outside until summoned to re-embark about an hour or so later. All I saw of Jakarta from our shed was the airport road; what impressed me most was that it was swarming with people, and that almost every woman of child-bearing age appeared to be either pregnant or carrying an infant on her back. At Darwin we stopped only to refuel, and went on overnight to Sydney where we touched down at Mascot airport at precisely the time stated in the Qantas timetable – 7.30 a.m. I have given a rather detailed account of the trip, partly because the air trip to Australia was then a very gentlemanly voyage, contrasting strongly with the rushed trip one has nowadays.

I fell in love with Australia at once, and, although I have visited it many times since, I still remember my first contact – the vastness, the uniform grey green of the vegetation, the strangeness of both flora and fauna, and the eerie silence of the bush. I liked the people, too, and got on very well with all types, and not merely with the academics. I moved around so much that I did not spend a great deal of time in the University of Sydney, and, although I delivered a few lectures, I did not give a formal course. R. J. W. Le Fevre, the head of the chemistry department, and his wife Cathie were extremely hospitable and helped make my stay a happy one, but he was having a difficult time. He had come to Sydney from London at the end of the war to inherit a department whose members were at loggerheads with one another, and, in some cases, resentful of Le Fevre's appointment. This sad state of internal strife seems to have dogged Sydney chemistry ever since I have known it, and it has, in my view, prevented what could easily have become a great school from realising its true potential; it has produced some really outstanding men like Cornforth and Birch, but they have all made their triumphs elsewhere. Even today, I have the impression that tension still exists there.

I visited all the capital cities and, with my hosts in each of them, saw quite a bit of the surrounding country, from the sugar cane fields and rain forests of Queensland to the desert, aflame with wild flowers, in Western Australia. I think that, on the whole, I was most attracted by Perth and Adelaide, and I was particularly impressed by the University of Adelaide which, unlike most of the others, seemed to have achieved some kind of rapport with the state government, whereas there appeared (to me at least) to be little sympathy between state and university elsewhere, although all of the universities were state organisations. In Adelaide I stayed with the professor of chemistry, that friendly, talkative Ulsterman, A. K. Macbeth, and spent quite a bit of time moving around the vineyards with that rumbustious character, Hedley Marston, who did so much for agriculture and the livestock industry in Australia, through his trace metal studies. He was something of a *bon viveur* – I recall that, when I stepped off the plane at the old Parrafield airstrip, he lurched towards the plane brandishing a walking stick and bellowing 'Look Alex – I've managed it at last – gout!' Dear old Hedley; he was always a thorn in somebody's flesh, but was, none the less, a great Australian. I still remember vividly trips through the bush country beyond Victor Harbour with Hedley and Mark Mitchell, the professor of biochemistry at Adelaide, where flocks of galahs used to swoop through the ghostly twisted gum trees, and leave behind them a bush which seemed even more silent than it was before. I learned a lot about Australia and its scientific and technical problems on these excursions.

In 1950 Australia was, in many ways, a fascinating country. It made me think of a giant who had for long been fast asleep, but was now beginning to stir before waking up. Architecturally and socially its cities were like transplants from pre-war Britain, although in politics and in labour relations they seemed to me much more raw and aggressive. It was an interesting time from an academic viewpoint, for the university scene was on the threshold of great changes. The creation of a university of technology in Sydney had been agreed, and some staff appointments had been made in 1949 although the new university (now the University of New South Wales) had as yet no buildings of its own, and, when I was there in 1950, most of its work was being carried on in the Sydney Technical College, which was (and, I believe, still is) located in Ultimo, not very far from the University of Sydney. Philip Baxter (now Sir Philip), the professor of chemical engineering in this new university, had taken up office about a year before I visited Sydney; I had known him in England, where he was Research Director at I.C.I. Ltd General Chemical Division at Widnes. We had been associated, partly through chemical defence work during the war, and subsequently in my capacity as consultant for several years to the General Chemical Division. He and his wife were very kind to me in Sydney, and I was greatly impressed by the way in which he was getting a grip on things, not only as regards chemical engineering in the new university, but through his development, even at that early stage, of good relations with the government of New South Wales, and the way he appeared to be taking a leading part in matters relating to atomic energy. At the time of which I write, the Vice-Chancellorship of the infant university was temporarily occupied by a civil servant from the New South Wales Department of Education, but it was clear that he was, in effect, just a stop-gap until the university had premises of its own. That, I confess, seemed to me a long time off. Philip Baxter took me one warm and sunny morning to see the university site at Kensington, where building was said to have begun. It consisted of a large open space with, somewhere near the middle, a brick wall about thirty feet long by ten feet high, evidently the beginning of a larger structure; at the base of the wall, on its shady side, reclined half a dozen builder's men apparently asleep. Such hives of activity seemed to me to be not uncommon on construction sites in Sydney, so I reckoned (erroneously, as it turned out) that the University of New South Wales would take a long time to develop. One thing was clear to me, however – Philip Baxter would surely take over as Vice-Chancellor; this he did, in due course, and the impressive new university is in effect a monument to his ability and drive.

In general the universities were in a poor way. They were swamped by elementary teaching and freqently at loggerheads with the state governments on which they depended for support. Research was at a rather low ebb; most of the best people they produced went abroad to study for a doctorate, and usually did not return. The situation, at least in chemistry, was not helped by the low level of industrial activity, with a consequential paucity of openings for graduates outside the institutes of the Commonwealth Scientific and Industrial Research Organisation, which were doing an excellent and very necessary job in supporting agriculture and livestock production on which the country's welfare mainly depended. Shortly before my visit, it had been decided, on the advice of some leading expatriate Australian scientists, spurred on by Sir David Rivett, the head of CSIRO, to create an Australian National University at Canberra – a decision which, not surprisingly, was greeted with no great enthusiasm by the existing state universities. I recall being at a dinner in Melbourne given, if I remember aright, by the board of I.C.I. (ANZ) Ltd at which Rivett and I were guests. He expounded his views on the national university scheme indicating that it was to be essentially a research university, where all the best Australians would be able to develop work which would achieve international recognition. I asked whether it might not be wiser at this stage to do something that would put the state universities in order and enable them to develop into world-class institutions, rather than embark on the national university scheme. With this view he disagreed vehemently, saying that the state universities were quite hopeless and could never develop in the way I suggested. In reply I said that, if the state universities were neglected, reliance on a national university would be no solution, and that, if the state universities were not properly supported and encouraged, the long-term outlook for science in Australia would be bleak indeed; to my mind, one should build up the state universities and create a national postgraduate university later on. We had quite a set-to that evening. In the event, of course, I was proved right in one respect by the development which has since occurred in the state universities, which are now flourishing both in teaching and research. On the other hand, as Rivett hoped, the national university – although I still think it was founded earlier than it should have been – has settled down and is now a fine institution.

A few days before I was due to leave Australia for home I received a telegram from my old friend and former Vice-Chancellor, Sir John Stopford, in his capacity as deputy chairman of the Nuffield Foundation. It informed me that I would find at home a formal invitation to become a Nuffield Trustee and finished with the words 'and don't dare to refuse'. So, on returning to England I became a Managing Trustee of the Nuffield Foundation. Thus began a long and intimate association with this great charitable foundation, which I have served successively as Managing Trustee, deputy chairman and chairman until the end of 1979, and since then as chairman of the Ordinary Trustees. The Nuffield Foundation was created by Lord Nuffield who, having made a large fortune in the motor industry, wanted to set up a trust which would apply his wealth to the advancement of health, and the prevention and relief of sickness, the advancement of social well-being by scientific research, the development of education, and the care and comfort of the aged poor. During the first twenty-five years of its existence the Foundation devoted most of its resources to the promotion of research by grants-in-aid to individuals or groups seeking to explore and develop new areas in science, medicine, or social studies. Through these, and through its large ventures in various fields, e.g. the development of radioastronomy at Jodrell Bank, the Nuffield Science Teaching Projects and its stimulation of some relatively neglected fields of science, I believe the Foundation made a contribution to science in the post-war years in this country and in the Commonwealth out of all proportion to the actual sums of money it provided. A detailed account of the Foundation and its work during its first twenty-five years of existence has been published in book form (Clark. *A Biography of the Nuffield Foundation*. Longmans, London, 1972), and I need not enlarge on it here. Suffice to say that I have enjoyed every minute of my association with the Foundation, and that I have learnt through it that to spend large sums of money wisely on education and research is far from being an easy task!

The year 1951 was an exciting one scientifically, because it was in the early summer of that year that we finally solved the problem of nucleic acid structure, and I announced it in a lecture given in Manhattan Centre, New York, on the occasion of the 75th Anniversary of the American Chemical Society in August. Attendance at that meeting was quite an experience, for I had never been at such a huge gathering before. We had a very pleasant trip from Southampton to New York on the Cunard liner *Caronia* in company with a number of chemical friends – Ewart (Tim) and Frances Jones (Oxford), Bill and Carol Dauben (Berkeley), John and Kathleen Lennard-Jones (Cambridge), and Vlado and Kamila Prelog (Zürich). We plunged into a veritable maelstrom of social and scientific activities, in the company of several thousand other participants. It is true that one did meet most of the world's leading chemists there, but actual encounters were fleeting, and I am afraid the whole thing convinced me that gigantic meetings of that nature were things to avoid wherever possible.

During the next few years I was kept pretty busy between my research in Cambridge, the work of the Advisory Council on Scientific Policy and overseas travel, for, probably as a result of our nucleotide work, I began to receive more and more demands to lecture abroad. So it was that I found myself in Lucknow in January 1953 speaking at the Indian Science Congress. This was my first visit to India, although, since India became independent, I had been under continuous pressure from my friend Sir Shanti Bhatnagar to come out and see the results of his efforts to develop the National Laboratories under the aegis of the Council for Scientific and Industrial Research (a government organisation rather like the Department of Scientific and Industrial Research in the U.K.). At Bhatnagar's request, following the Lucknow meeting I spent several weeks visiting in turn Allahabad, Benares, Calcutta, Madras, Poona, Bombay and Delhi. I would find it very hard to give any simple or straightforward description of my impression of India on that visit. India had so many relics of a glorious past, and yet it seemed that all the glory was indeed in the past, and that what remained was a vast heterogeneous country in which one saw poverty such as I had never even dreamed of, cheek by jowl with fantastic riches. It was a land of contrasts, and I think my principal reaction was one of uneasy fear in face of a situation which was, to say the least, potentially explosive, and perhaps would actually have been so, if the general population had been better nourished and so lost its dull apathy. And I must confess that this same uneasy fear is with me even today, despite (or perhaps because of) a number of subsequent visits. This feeling I found most marked when in central and northern India; the south I found much more attractive, and it seemed less poverty stricken although the extremes were still there. I recall one remarkable encounter with the wealthy side of India when I was visiting Madras.

As I indicated above, Shanti Bhatnagar was very busy creating the National Laboratories, and I was in Madras when the National Electrochemical Laboratory was about to be opened in a small and rather (to me at least) obscure place called Karaikudi down in the southern tip of the country. I confess that, when I heard of it, I wondered why on earth one would want to locate a laboratory in such a remote spot. I learned, however, that a wealthy local landowner (named

Alegappa Chettiar) had offered to pay for the entire building if it were located in Karaikudi, and this offer had been accepted by the government. For the opening ceremony, which was to be performed by the Prime Minister, Alegappa Chettiar had a large swathe cut in the jungle near his home, and tidied it up so that large aeroplanes could land. So we all flew down from Madras in two specially chartered Air India Constellations, had lunch with Chettiar, then went through the opening ceremony and flew back to Madras. Lunch was served to somewhere between fifty and a hundred people, and, as far as I could see, most of the plates and goblets used were silver or gold. I was given to understand that everything was paid for by Chettiar; if so, it must have set him back quite a bit, although he seemed wholly unconcerned.

I also attended in Madras the opening of the National Leather Research Laboratory – a hilarious afternoon. The opening ceremony was performed by Sir C. V. Raman, who devoted his remarks to the iniquity of cow slaughter, and claimed that the feet and ankles of pretty girls should be visible and not covered up by leather (or indeed anything else)! I thought the Director of the new laboratory was going to have a stroke, but he managed to restrict himself to a blistering attack on Raman who did not, I fear, take it at all kindly. To complete the afternoon the lights fused when the platform party was viewing the tannery, and its members had to stand in the dark for about ten minutes, not daring to move on the narrow walkways they were traversing between the somewhat malodrous tanning pits.

When I got to Delhi at the end of my trip, Bhatnagar was anxious that I should talk with Pandit Nehru. This, it appeared, could not be arranged until the day after I was supposed to leave for London. After much fussing I was transferred to a flight two days later on the Comet, which was then the great novelty in air transportation. (Incidentally I did fly home on it, and found it an exciting experience and a portent of things to come. I would perhaps have enjoyed it less had I known that, only a week or so later, the plane I flew in was to disintegrate over the Mediterranean because of metal fatigue!) I went to the Prime Minister's house for lunch and found him somewhat distraught on my arrival. He told me he had had a bad morning. First of all he had received the Persian leader Mossadeq, who, he said, was very difficult to deal with since, whenever one said anything he didn't like, Mossadeq would retire to a corner and weep. Secondly, when Mossadeq had gone Nehru had returned to his study to find that a monkey had got in through an open window and scattered his papers all over the room. However, he quickly recovered his good humour, and we had an interesting talk about India and what science and technology might do for it. His daughter, Indira Gandhi, was hostess and I found her very impressive; indeed from that day onwards I never had much doubt about who would succeed Nehru when the time came.

Shanti Bhatnagar had a cousin, Colonel S. S. Bhatnagar, a well-known medical man in Bombay, and, through him, on this trip I met a number of people who became fast friends. In particular I think of Dr K. A. Hamied, the owner of a flourishing pharmaceutical company, Cipla Ltd. He was a remarkable man. A Muslim from the Central Provinces, he had taken his chemical degree at the Muslim University in Aligarh and then gone to Berlin for his doctorate. There he met his wife, a Polish girl, and, after they were married, he returned to a university lectureship at Aligarh. He very soon found that they simply could not live on the pittance he was paid, so he resigned and, with his wife, set off for Bombay and settled into a hovel in the outskirts of the city (just as many people still do today). At this point he possessed 100 rupees, and, with them, he started to make and sell love philtres, indigestion cures, aphrodisiacs and so on. Out of this grew his company, and he became in due course a wealthy man. He was a member of the Congress Party and a follower of Gandhi, and had several spells in prison in the years before independence. When independence did come he elected to stay in Bombay although a Muslim; he did not have any trouble on that account, and was, indeed, Sheriff of Bombay for some years. He was, as I have said, a remarkable man and I enjoyed his friendship until his sudden death some years ago. In 1953 he introduced me to his son Yusuf whom he was determined to send to Cambridge to study chemistry under me. This Yusuf did, and did it well, with a first-class degree and a Ph.D. He has the same drive and entrepreneurial quality as his father in addition to being a first-class chemist, and, under him, Cipla has gone from strength to strength as an ethical pharmaceutical company with its own research and development department.

In the following year, 1954, shortly after receiving a knighthood in the Birthday Honours, I spent the Fall term (September–December) at the Massachusetts Institute of Technology as Arthur D. Little Visiting Professor of Chemistry and gave a course of lectures on vitamins, coenzymes and nucleic acids. These, for some reason, were a great success and, twice weekly, my audience included most of the chemical and biochemical staff, research students and senior undergraduates from both M.I.T. and Harvard. A by-product of this was a minor flood of American postgraduate students in subsequent years in my laboratories in Cambridge. It was a most enjoyable visit; my wife Alison was able to spend a substantial part of the term with me, and we rented a service flat in the old Hotel Continental in Cambridge not far from Harvard Square; there we could, and did, entertain our numerous friends – the Woodwards, Bartletts, Blochs, Buchanans, Westheimers, Sheehans, Copes and also the W. S. Johnsons (who were on sabbatical leave from Wisconsin). At the end of our stay we had a hilarious farewell party; photographic and other records of that and other parties of the period are still highly prized possessions of a number of the participants. I saw a lot of R. B. Woodward during this stay, and our friendship became even closer. As I have already mentioned, Bob was a remarkable man with a devotion to organic chemistry I have never seen equalled by anyone else, coupled with a prodigious memory, an enormous capacity for hard work, and more than a streak of genius. In those days he worked long hours in the laboratory, and his research seminars were already famous. He used to hold them in his room in the Converse Laboratory at Harvard late in the evening, and continue them into the small hours. It was during these seminars that many of his brilliant ideas were advanced, apparently 'off the top of his head'. Needless to say, not all of them were; they were, more often than not, the result of exhaustive reading and study, but it was one of his characteristics that he liked to adopt the pose of a genius who plucked ideas out of the air. The same affectation could be observed in his lectures. He was no mean actor, and his famous little box of coloured chalks with which he meticulously drew chemical formulae on the blackboard (in those days he never used slides), and the absence of any lecture notes despite the inordinate length of his lectures, all helped to conceal from his audiences the amount of hard work he put into lecture preparation. But this foible did not detract from his brilliance, nor did it conceal from those of us who knew him the loyal and generous friend behind it. Woodward was, I believe, the greatest organic chemist of his generation, and his sudden death in 1979 was a great loss to science.

Shortly before the end of our stay at M.I.T. I had a message from the Colonial Products Council in London (of which I was a member) asking me to go down to Trinidad on my way home, to look at some problems in their cocoa research establishment, and, incidentally, to visit the sugar research laboratory at the Imperial College of Tropical Agriculture. We left New York at a temperature of -5 °C and found ourselves a few hours later in Trinidad at $+30$ °C – a rather trying transition dressed as we were in clothing more appropriate to the former. We lodged at the Imperial College at St Augustine with Dr Herklots, the Director, and his wife, and spent an enjoyable few days discharging the tasks allotted to me (which were hardly arduous). Trinidad was an entirely different kind of place from anything I had seen before – a tropical colonial possession with a multiracial population much given to laughter, and to singing and dancing to the music of steel bands; it also possessed some of the most reckless taxi-drivers I have ever encountered.

By the end of 1954 our collaborative work on the structure of vitamin B_{12} with Dorothy Hodgkin was well advanced. Work on the same subject was, of course, being intensively pursued by Karl Folkers and his group at Merck & Co. Inc. at Rahway,

New Jersey, and, while I was at M.I.T., I was invited to visit the Merck research laboratories to talk informally about vitamin B_{12}; it turned out to be rather heavy going, for it very quickly became evident that although the Merck group wanted to learn all I knew, ideas of commercial secrecy made them determined to give me no information of any value about their own work. Needless to say, this attitude caused me to adopt a similar stance, and so we succeeded in getting nowhere. Although it was frustrating at the time, it seems rather ridiculous in retrospect. We – or perhaps better said Dorothy – finally fixed the structure of B_{12} in the summer of 1955. That year the International Union of Pure and Applied Chemistry was holding a Congress in Zürich, at which I was present. The Congress president was Paul Karrer, Professor of Organic Chemistry at the University of Zürich, a man well known to be at daggers drawn with Leopold Ruzicka, his opposite number at the Eidgenossische Technische Hochschule (E.T.H.) also in Zürich. The B_{12} structure had been completed too late for inclusion in the Congress programme – my recollection is that we published it in *Nature* during or just after the Zürich meeting. I soon found myself in a difficult situation. Both Ruzicka and Karrer were my friends and I found myself rather delicately placed; Ruzicka wanted me to speak at the E.T.H. on B_{12} at the same time as a major Congress lecture was to be given at the University on the other side of the road. Ruzicka's idea was that everybody would flock to B_{12}, and leave the other lecture with Karrer in the chair sadly depleted. I had to be firm about it, and, although I did in fact give the first public presentation of the B_{12} story in the E.T.H., I did so after the formal Congress lectures were finished and so avoided a real rumpus.

In, I think, early autumn 1955, a group of Soviet ministers for a variety of industries, including the chemical industries, led by A. N. Kosygin, then Deputy Chairman of the Council of Ministers, visited Britain. The party went around the country seeing various industrial operations, and, at the end of their trip, they had a week-end in which to do sightseeing in London and pay a visit to Cambridge on the Sunday as guests of the university. The Vice-Chancellor at that time, Professor B. W. Downs, was also Master of my own college (Christ's), and he invited me to join the party for lunch in Christ's, largely on the grounds that he, a Scandinavian languages expert, was somewhat alarmed at having to entertain a group of technocrats. When I arrived for lunch, the Vice-Chancellor said he had learned to his horror that he was supposed to look after the party until tea-time and he had made no arrangements for such an eventuality. Could I find some way of looking after the visitors for him?

When I came to Cambridge in 1944 one of my conditions was that the university would give top priority to a new building for chemistry. The university was as good as its word, although, what with licensing problems and steel shortages we were unable to make a start until after 1950, and even then had to proceed very slowly. However, by the time of the Soviet ministers' visit it was approaching completion, and so I suggested to them that they might like to see the new laboratory. They said they would very much like to do so, so off we went and made a tour of it. At the end of it Kosygin said they had much enjoyed seeing it, but presumably we had another laboratory which was currently in use – could they see it? I said 'Certainly,' and we all trooped off to the old chemical laboratory in Pembroke Street. In the course of walking around it, we passed through one of the research laboratories in which four young men were busily working, whereupon Kosygin remarked 'I see that you make your students work on Sundays.'

'Not at all. These men are not being made to work. They are, in fact, postdoctoral research workers who are here on Sunday afternoon of their own free will, because they want to get on with their research.'

Kosygin looked a bit doubtful so I added 'Would you like to meet them?'

'Yes,' was the answer, so I called them over, introduced them and had them tell Kosygin what they were doing. It so happened that the group comprised an American, an Australian, a New Zealander and a Scotsman.

'You seem to have a very international group,' said Kosygin.

'Yes,' I said, 'I should think that about half of my research school comes from overseas.'

'That is very interesting, for you must have formed some opinion of the training of chemists in various countries. Where do you get the best people – from Germany perhaps?'

'No,' I said, 'my experience is that, in general, the German students are trained to do just what they are told by their professors, and are not encouraged to be original.'

'Oh! – then perhaps the Swiss – they have strong chemical industry too.'

'No, they are rather like the Germans in their attitude.'

'Well then, what kind do you like, and where do you get them?'

'I like them to display originality in approach, and that I find most often in Australians, Americans, New Zealanders and our own British students. Unfortunately, I can't really say anything about Soviet students.'

'Oh! Why not?'

'You know perfectly well why not. You never let them out.'

'Yes, we do!'

'Yes, to Bulgaria or Poland, etc. You don't send any of them here.'

'You wouldn't take them.'

'Who said I wouldn't take them? I couldn't take a hundred but I give you my word here and now that I am willing to take two of your young researchers whenever you care to send them to me.'

'Do you mean you will take our students?'

'Certainly. But I have three conditions – (1) they must be good chemists or I'll send them home to Moscow; (2) this is a chemical laboratory containing people from a variety of countries. We have nothing to do with politics, and your people must come here, live and work with the others in the laboratory, leaving politics alone; and (3) I don't want to have them under continuous supervision by your Embassy officials.'

'Fair enough,' said Kosygin. 'I'll think about this when I get back to Moscow.'

And we left it at that.

About a couple of months later a Soviet Academy of Sciences delegation led by the President, A. N. Nesmeyanov, came over to Britain as guests of the Royal Society, and I was asked by our President, Lord Adrian, to join him and a few others at a sherry party given by the Royal Society at the Athenæum in London to welcome the Russian party. When the Russians arrived Nesmeyanov promptly got hold of me and said 'Todd, I want to talk to you.'

'Certainly,' I said, 'why not now?'

'Fine!' said Nesmeyanov. 'Kosygin has been to see me and told me that you would accept a couple of Russian research students. Is this true?'

'Yes – but subject to my conditions.'

'I know – Kosygin told me your conditions and we accept them. When can you take the two men?'

'Anytime you wish.'

'All right, I'll send them to you in September 1956 and they can stay with you till April 1957.'

And so it was. In September the first Russian research workers (both postdoctorals) to come to this country after the war – N. K. Kochetkov and E. A. Mistriukov – came and worked in Cambridge. The experiment was most successful; the two young Russians fitted in well, and were popular members of the laboratory. They became and remain my close friends; Kochetkov, now an Academician, is Director of the Zelinsky Institute of Organic Chemistry in Moscow, and Mistriukov is a staff member of the same Institute.

In this way contact with Russian research was re-established, and since then a two-way exchange traffic in research workers between the Soviet Union and this country has developed much, I believe, to our mutual benefit. An interesting sidelight on this matter of scientific exchanges with the Soviet Union prior to my intervention is cast by a conversation I had with Nesmeyanov not long afterwards. He asked me how I had managed to arrange it with the British authorities to get the young Russians admitted, because they had tried for some years through scientists known to be friends of the Soviet

Union (e.g. Blackett and Bernal) to get exchanges started, and had failed completely. I pointed out that, bearing in mind the somewhat uneasy relations between our two countries over nuclear matters in those days, there would seem to be little hope of success if he and his colleagues tried to set up things *via* physicists with well known left-wing affiliations! Had they simply approached the Royal Society, matters would probably have been quite different.

Shortly before our Russian students left for home, my wife and I were invited to visit the Soviet Union as guests of the Academy of Sciences, and, accordingly, we made our first visit to that country at the end of May 1957. A visit to Russia in those days still had an exotic flavour, but was not without its problems. We could get little or no information about our trip, received visas only at the last moment and set off from London by Finnair not knowing where we should stay or who, if anyone, was to look after us on arrival in Moscow. As it turned out, of course, we needn't have worried. We were greeted on the tarmac at Vnukovo airport by a large Academy group bearing masses of flowers for my wife. All formalities were attended to by an official on our behalf, and we were lodged in a comfortable Edwardian style suite in the Hotel Moskva in the centre of the city near the Kremlin and the Bolshoi Theatre. I lectured at the Zelinsky Institute, and we had the full treatment of sightseeing, opera and ballet. Our visit coincided with one by Sir Malcolm Sargent, who lodged in the British Embassy with our ambassador Sir Patrick Reilly and his wife. All of us visited Leningrad together, and attended there a concert conducted by Sir Malcolm and enthusiastically received by a large audience. The Russian visit was very interesting, and we made a number of new friends, but Moscow, and even Leningrad, despite its elegant buildings and the Hermitage Museum, struck me as drab and with a rather oppressive atmosphere. The drabness was emphasised by the obvious shortage of consumer goods and the indifferent quality of those that were available, and by the monotonous appearance of the rather poorly constructed box-like buildings which had been springing up all over Moscow and Leningrad since the war, to try to cope with the housing problem, which had arisen, partly from the war, and partly as a result of the movement of population from the countryside to the cities. I felt it oppressive, because we seemed to be under continuous surveillance; indeed, only during a day's picnic with several friends at Academician Nesmeyanov's *dacha* or country house outside Moscow, did we feel ourselves in really relaxed surroundings. Even when we went to Sochi on the Black Sea coast for a few days with our guide-interpreter, I had the same feeling of regimentation, which only left when we were on a brief excursion into northern Georgia. It is only fair to say that, in numerous visits I have since paid to the Soviet Union, matters in these respects have been steadily improving, although I feel that, even today, the Russians accept, as a matter of course, a degree of regimentation through their all-pervading bureaucracy, which most people in England would find hard to accept. One beneficial effect of this first visit to Moscow was that it caused me to do something about my Russian. I had learnt a bit of Russian many years before, largely because it seemed a rather interesting language, but, in the years that followed, I had never had occasion to use it. As a result, when I was in Moscow in 1957, I found, to my dismay, that I had almost wholly forgotten it, although, after a few days, I began to recall odd words and phrases. This experience caused me, on my return to England, not to start formal lessons but to persuade a friend, Mrs Natasha Squire, who taught Russian in the university, to submit to the trial of holding (or trying to hold) regular conversations with me. Slowly this worked, and over a few years I recovered much of what I had lost; this greatly added to my enjoyment of subsequent visits to the Soviet Union.

7

The Nobel Prize and its aftermath

—

The year 1957 was, of course, a red-letter one for me, since, in the autumn, I was awarded the Nobel Prize for chemistry. The award was a great surprise to me, for I had never given any serious thought to such a possibility. It is true that, when I was visiting Adelaide in 1950, A. K. Macbeth had said to me one day that, if some of the things I was attempting in the coenzyme field came off, I could well be a candidate for the Nobel Prize. I am afraid I reckoned that his comment said more for his heart than his head, and I thought no more about it. In the latter part of September and the first week or so of October 1957, however, I was in Berkeley at the University of California as a Visiting Professor delivering the Hitchcock Lectures. There I spent a good deal of time with my friend Wendell Stanley, one of America's great figures in biochemistry (and, incidentally, my mentor in the intricacies of American football, a game which we both enjoyed). One evening towards the end of my stay we were having dinner together, when Wendell raised his glass and said 'Your very good health – and remember this toast in December.' I confess that I couldn't think what he meant and, since he clearly didn't want to enlarge upon it, I did not seek to probe the matter. I had promised to spend a few days with Bob Woodward at Harvard on my way back to England, so I flew to New York when my lectures were finished, met Bob there, and with him set off for Boston, as we often did, by rail on a New York Central train named the *Yankee Clipper*. As we bowled northwards, consuming a succession of drinks in the club car, the subject somehow got around to possible candidates for the Nobel Prize, the award of which was likely to be announced about the end of October. We discussed various names, and then Bob said 'Of course there are also two rank outsiders in the field – you and me!' 'There could be something in that as far as you are concerned in view of your cholesterol and strychnine syntheses,' said I, 'but I would lay very long odds against me.' So we had another drink and left it at that.

It had been arranged that I should stay with the Woodwards at their home in Belmont and we arrived there shortly before dinner. To my surprise Doxie (Bob's wife) handed me a letter which had arrived that day from my wife; I had not expected one, since it is always a bit chancy sending letters across the Atlantic if the intended recipient is moving around and stopping for only short periods in one place. I had a look at the letter as I went upstairs to tidy up before dinner; the main point in it was that Robert Robinson had asked Alison for a photograph of me which was wanted for the Swedish press. I would have been peculiarly dense if I hadn't realised the significance of that request, but I still remember the odd feeling I had at dinner, remembering the discussion on the *Yankee Clipper* and yet unable to say anything about it. After the award was public, and often thereafter, Bob Woodward and I used to recall the incident with amusement. The actual announcement of the award did not come for about a fortnight or so after that night in Belmont, by which time I was back home in Cambridge, and it was not altogether straightforward and simple. The first indication was when a journalist of the *Dagens Nyheter* of Stockholm descended upon me to congratulate me on winning the Nobel Prize for chemistry, and wanting an interview; this was closely followed by the international press agencies and the British press. In vain I protested that I had received no notification, and indeed I did not give any interview until after I had received – a couple of days later – an official telegram from the Swedish Academy.

That, to the best of my recollection, is the full story of the way in which it all happened. I did later make some discreet enquiries to find out how Wendell Stanley had been so sure of his ground; it transpired that I had emerged as the only contender for that year's prize quite early in the Committee's deliberations, and that he was a close friend of the chairman! The actual presentation ceremony in Stockholm and the accompanying festivities have often been described, and, from the prizewinner's standpoint, are quite marvellous. Particularly impressive, I found, was the way in which, from the moment our plane touched down in Stockholm until we left Sweden, my family and I were accorded, quite literally, the kind of treatment normally given only to royalty or heads of states. One comes down to earth with a bump when one alights at London airport on return and is put through all the usual tiresome formalities with no consideration at all. I remember how, on our return, a rather officious young customs officer at Heathrow made me open up my suitcase and rummaging in it he came across my Nobel Medal. 'What's this?' he said. 'Is it gold?'

'Yes,' I said.

'Then you will have to pay quite heavy duty on it.'

It was clear that he actually meant it, so I said 'I really wouldn't advise you to charge duty on that. Do you know what it is?'

'No, but what has that got to do with it?'

'Well, it's the Nobel Prize Medal and you are going to look very foolish in the press tomorrow, when it is reported that you charged duty on it. If I were you, I think I would let it through.'

And he did!

On the occasion of the dinner given by the King and Queen of Sweden my partner was Princess Margarethe – a tall, strikingly beautiful young woman and a good conversationalist. At about this time, there was quite a lot of speculation in the gossip columns of the international press about who she might marry, and who she was with at this or that function. Naturally, photographers are everywhere during the Nobel celebrations and one of their shots appeared, showing Princess Margarethe and myself at the dinner table apparently having an animated conversation, and she with her left arm slightly raised, as though she were emphasising a point. After I got back to England I had a telephone call from a member of the staff of one of the London Sunday newspapers, who said that he had noticed that the Princess appeared to have what looked like an engagement ring on the third finger of her left hand, and perhaps I could tell him about our conversation – did she say anything about it, or about X (a person with whom her name had recently been coupled by the gossip columnists)? I almost exploded, and I remember telling him that, not only would I tell him nothing about any conversation I had had, but I took it as an insult that he should even have thought that I might! I did, however, remind him that in Europe engagement rings are worn on the right and not on the left hand! As a result of this and one or two other subsequent experiences with journalists of the yellow press, I find myself at times wondering where they are recruited.

There has been a lot of nonsense both talked and written about Nobel Prizes, and the alleged striving for them by candidates in fierce competition with each other. One of the most striking examples is, in my opinion, an appallingly bad novel called *The Prize*. I have known many Nobel prizewinners, and I have yet to meet one who would conform to the kind of picture given there. Most are people who, like myself, were surprised and deeply honoured by the award, and only a few could be said to have set their sights on the prize and worked, as in a competition, to win it. For a scientist to set his heart on winning a Nobel Prize is, in my view, foolish. After all, if we consider any one subject – for example chemistry – there must be, in any year, a considerable number of candidates any one of whom would be worthy of an award, and, as a rule, there will be a lottery-like element in the final selection. There must be many chemists in the world whose work would warrant an award and who have not received one. It is to the credit of the Nobel Foundation, however, that few, if any, have been honoured whom their scientific colleagues would regard as unworthy to receive a prize. More than that one could not ask of any selection committee.

The award of a Nobel Prize brings to the recipient a great

many things which can, for a time at least, grievously interfere with his research. He is suddenly in great demand as a lecturer in every corner of the globe, is showered with honorary degrees and deluged by requests from autograph-hunters. He suddenly finds that he has become a kind of oracle, who should be listened to with bated breath on every subject under the sun. In particular, he will receive a stream of requests to append his signature to public declarations or letters of protest about humanitarian or, very frequently, political problems. It is curious how some people seem to feel that winning a Nobel Prize in, say, chemistry suddenly makes their opinion on wholly non-scientific questions more significant than those of other people within whose province they really lie. But when all is said and done, it is a wonderful thing to receive the accolade from one's fellow-scientists, and perhaps prizewinners display no more than the normal failings of their fellow-men.

The building of the new laboratories, which I had been promised on coming to Cambridge, took a long time, partly through vacillation by the university about the site, and more especially because of governmental restrictions. The latter were, of course, not unreasonable, as so much war damage had to be repaired, but the former was rather tiresome. In the end, however, J. T. Saunders (Secretary General at the time of my arrival) and I had our way, and it was agreed that we should build on the site of Lensfield House, an old mansion with a large garden on the south side of Lensfield Road. Saunders and I favoured this site, not just because it was big enough to accommodate what was to be the largest chemical laboratory in Britain, but because, abutting it on the south and stretching as far as the University Botanic Garden was a rather decayed residential area known as New Town, which was ripe for demolition and rebuilding. It was clear to us that expansion of the physical sciences and engineering in Cambridge was bound to happen, and that New Town provided an ideal building site which would keep all the sciences together on the right bank of the Cam, near the colleges, with the humanities continuing to develop on the left bank. Policies of universities change with changing administrative officials, however, and despite protestations by myself and others, this development scheme was not followed, and, some ten years or so after chemistry had been completed, the decision (in my opinion a wrong one) was taken to build physics anew on a remote site in west Cambridge; piecemeal development of this type is, in my view, undesirable and could prove very damaging to science in Cambridge.

As architects for the new chemical laboratories we chose the firm of Easton and Robertson. Murray Easton and his junior partner, Teddy Cusdin, had done a lot of hospital building, and were keen to build a really modern laboratory. At this time the University Grants Committee, realising that there would be a lot of building to be done in the late fifties and sixties, was keen to see some more or less standard type of construction devised for scientific laboratories, and its chairman, Sir Keith Murray, gave me a free hand to experiment. Easton and Cusdin, aided considerably by Ralph Gilson whose knowledge of the operation of laboratories was unequalled, rose to the occasion. We built on a steel space frame so that, on each floor of the building we needed no supporting walls, and services were arranged so that complete flexibility as regards laboratory or other uses could be achieved by using easily demountable partitions and building on an eight-foot module system. We began in 1950, but progress was painfully slow because of steel rationing; we used to get small deliveries at rather long intervals, and so could only do actual building work for short periods at a time. This, as it turned out, had one great advantage; we decided to build a full-scale, three-module section of our proposed building on one corner of the site, and in it we tried out all our fittings, and planned the most convenient and durable laboratories we could think of. In this we were aided by my research group, members of which used to go to the model laboratory, and not only check up on convenience of operation but on its destructability. They poured masses of organic solvents over floors and benches, set the lab on fire on occasion, and tested out shelving by overloading to the point where it tore away from the walls. When we were finished with the model and had decided on the best materials and design, we simply told the architects to repeat the pattern throughout the new laboratories. The method certainly worked, for not only are the laboratories still in excellent shape after some twenty-five years in use, but their main features have been incorporated in many chemical laboratories built subsequently in many parts of the world.

With the completion of the first phase of the new Lensfield Road laboratories incorporating organic, inorganic and theoretical chemistry in 1955–6, Ralph Gilson and I felt we had achieved what we had set out to do together in Cambridge. It was abundantly clear that he was a man of such remarkable ability that to remain in Cambridge in the position he held there would be a serious waste of talent. When, therefore, he was approached by Richard Perkin, the head of the American scientific instrument firm of Perkin Elmer Inc. to leave Cambridge and become chairman of his British subsidiary Perkin Elmer Ltd, I encouraged him to go, greatly as I regretted the break-up of a happy and close association going back for nearly twenty years. Gilson has made a great success of the British Perkin Elmer, and through it a real contribution to scientific instrumentation in this country.

When Ralph Gilson left I could have been in serious trouble, for men who can do his kind of job well are not easy to come by, and I still had the completion of the final stages of the laboratories before me. In this situation I remembered Ron Purchase, a young English airman – an R.A.F. flight engineer – whom I had encountered first in Göttingen in 1946, when he was busily (and successfully) engaged with Bertie Blount in taking over the Aerodynamische Versuchsanstalt there and installing in it the collected human relics of the old Kaiser Wilhelm Gesellschaft. He impressed me then, as he did later when in 1950 I met him again in Canberra. He had gone there to join Sir Mark Oliphant who was building a physics complex for the newly emerging Australian National University, and was doing a splendid job. I guessed that the Australian building might now be less demanding, and that Purchase might be a little restless so I wrote him and offered him Gilson's job. To my delight he accepted, and he took over in Cambridge in the winter of 1956–7 following the end of the Australian academic year. It wasn't very long, however, before Purchase found himself getting involved in work of an unexpected character in addition to looking after the Lensfield laboratories.

In the 1950s there was (as there still is today) a lot of talk about the apparent backwardness of much of British industry in exploiting new science-based technology, the need to produce more highly trained and inventive engineers, and to have them not only revitalise industry but also bring the universities closer to the practical needs of the economy. Among industrialists and politicians the virtues of America in this respect were extolled and the Massachusetts Institute of Technology and the California Institute of Technology were frequently singled out as shining examples of what they would like to see in Britain. I need hardly add that most of them had only the dimmest idea of what these two institutions were and, as one who had been for a short time a professor in both of them, I had, at times, some difficulty in recognising them from the descriptions which were in circulation at the time. I knew, although only in general terms, that a group of industrialists, with John Oriel of Shell Ltd as a moving spirit, had been actively trying to set up some kind of technological institution but without a great deal of success. By 1956 I knew that their first idea of building on the base of Cranfield Aeronautical College had been abandoned, and that a second scheme, aiming to build an institute in association with Birmingham University, was also about to collapse. I was not surprised at the failure of these two attempts, nor was I greatly in sympathy with the scheme which had been proposed. I do not believe that you can build a replica of an institution of higher education which has grown up in one country, set it down in another with a totally different educational system, and expect it to succeed. Success will only come through the development of an institution in tune with the existing educational and cultural framework of the country in which it is located.

In December 1956 I had a note from John Colville (universally known as Jock), Sir Winston Churchill's private secretary, asking me to lunch with him at his office in London.

At this meeting he told me that he, Sir Winston and Lord Cherwell, had often discussed the problem of getting new technology into industry and the possibility of creating an institution like M.I.T. The same idea had been put forward to Sir Winston *via* Colville by Carl Gilbert of Boston, President of the Gillette Corporation. I was asked what I thought about it, and I gave the views I have just outlined. I pointed out that it was the people in an institution like M.I.T. that mattered, and that one of our troubles here was that the universities and colleges with special interests in technology, by and large, lacked the prestige to attract people of the highest quality either as teachers or students. In this country, it was quite clear that the prestige of Oxford and Cambridge ensured that the best brains in the country tended to gravitate there; this led me to suggest that one might do worse than take advantage of the collegiate system in Cambridge, where there was already a large academic engineering school; with money from industry, one could found a new college which would have a majority of its students in science and technology, would help develop strong research schools in the various technological studies, and in various other ways forge strong links with industry. After some further discussions, which also involved Mr. John Oriel of Shell Ltd and Sir John Cockcroft whom I brought in, it was agreed that the scheme should be closely examined. During the first part of 1957 I introduced Jock Colville to Professor B. W. Downs, Master of Christ's College and, at the time, Vice-Chancellor of the University of Cambridge, and to Lord Adrian, then Master of Trinity and due to succeed Downs as Vice-Chancellor in October 1957. Although not himself a scientist, Downs was an enthusiastic supporter of the idea, and he was mainly responsible for persuading the university formally to accept the idea of a new Churchill College operating on the above lines and forming a national memorial to Sir Winston. There was, of course, some opposition from diehards in the university, who viewed anything scientific with deep distrust, but this proved rather weak when put to the test and, after the usual series of academic manoeuvres, the outline scheme for the proposed new college was accepted by the

Council of the Senate in November 1957. Thereupon began a lengthy series of exploratory talks with industrial leaders in which Jock Colville and I were much involved. Our idea was that we should be privately assured that industry was likely to contribute £3½ million before we formally launched an appeal. In this we were greatly helped by Viscount Knollys of Vickers Ltd, who agreed not only to help in the exploratory work, but to become chairman of an Appeal Committee for the project. A Trust Deed was now drawn up and executed. In it the following were named as Trustees: Sir Winston Churchill, Lord Tedder, Lord Adrian, Viscount Chandos, Sir John Cockcroft, Professor Downs, Lord Fleck, Lord Godber, Sir Alexander Todd, Lord Weeks. Later Noel Annan, Sir William Carron, Viscount Knollys and John Colville joined as additional Trustees.

Once we reached the stage of approaching potential donors informally, I found myself saddled with the job of keeping track of amounts promised, and dealing with the local administrative problems such as opening bank accounts, etc.; my new laboratory superintendent, Ron Purchase, took on the job of acting as a Cambridge secretary to the Trust. In this work he continued until he left to return to Australia at the end of 1958. He did a first-class job, not only for the Trust but for me in the laboratory, and I was sorry to see him go; but I fear Purchase was an incurably restless individual who had to be always on the move. I have lost touch with him in recent years, but I used to hear from him occasionally for about ten years after we parted, and he seemed every time to be in different jobs located widely apart. He was, for a time, back in this country at Harwell, then later in Nauru in the Pacific – very much a rolling stone. I was doubtless lucky to have him when Ralph Gilson left Cambridge, but I was even luckier when I had to find a replacement. When I went to Manchester in 1938 Professor (later Sir) Ian Heilbron took with him to London his laboratory steward, F. G. Consterdine, a really first-class man under whom, incidentally, Ralph Gilson had been trained. Consterdine remained attached to Heilbron when the latter left Imperial College and went to direct the new Brewing Industry Research Foundation. When Heilbron died suddenly in 1959,

Consterdine did not find the new regime to his liking, and, hearing that I was looking for someone like him to replace Purchase, he agreed to come to Cambridge. He took over without fuss or bother and ran my department with quiet efficiency until his retirement some years later; the wheel indeed had come effectively full circle.

It is not my intention to give a detailed history of the development of Churchill College, of which I later had the honour to be made an Honorary Fellow, but I treasure many memories of those early days when the idea was first elaborated, the money collected and the building and staffing of the new college was undertaken. During the period between the signing of the Trust Deed and the formal opening of the college the Trustees used to meet regularly at Sir Winston's house in Hyde Park Gate, London, S.W. The meetings were great fun, although we did also discharge a lot of business. The hour was always, as I recall, five o'clock in the afternoon, and each of us was provided with an enormous cigar and a very generous portion of brandy, ample replacements of both being scattered around the table. We always arranged that Oliver Chandos should sit immediately to the right of Sir Winston at the head of the table. This was important, because Chandos had by far the loudest voice in the room, and Sir Winston had a way of removing his hearing aid when the meeting started and was liable, in Chandos' absence, to become *incommunicado*. Nevertheless, he used to enliven the meetings from time to time by turning on a stream of epigrammatic remarks on features of the day's business which took his fancy. And many of his comments were not only acute, but clearly showed that he took the whole college project very seriously indeed; he was no sleeping partner despite his advanced years.

In the middle of one of our meetings I remember Sir Winston fumbled in his coat pocket and produced a letter which he opened and then announced 'I have a letter here from a woman – a very interesting letter. She says we ought to have women in the college. Seems quite a good idea – why don't we have women in the college?'

Whereat Chandos roared in best parade ground style 'If you

do that, Winston, you might have to give back some of the money we've already collected.'

'Ah,' said the great man, 'that's different,' put the letter back in his pocket and replenished his glass.

On another occasion, when plans for the college buildings were being considered the question of a chapel was raised. Because we hoped the membership would be very international some of us felt that a chapel specifically associated with one religion would not be appropriate. The matter was speedily settled by Sir Winston who said 'Chapel? Why should we have a chapel – Cambridge is full of churches already.' But it didn't quite work out like that. Someone must have leaked this discussion in Cambridge, for, very soon thereafter, we received a small cheque – I think it was for £5 or £10 – as a donation for the chapel. It was felt that to return the cheque would probably lead to trouble, so we decided that, if enough money to build a chapel came to us earmarked for this purpose, then we would include one, subject to the proviso that it should be so designed that any member of the college would feel able to use it for private devotion, no matter what his particular religious beliefs. Such a chapel was in due course included in the final plans.

There was quite a discussion at one point about a motto for the college, and it was decided that between two meetings a paper should be sent to all the Trustees, on which they were asked to write their proposal for a motto. As it went to the Trustees in alphabetical order, I was able to see the proposals of all the rest except Weeks before adding my own. At the outset I was barren of ideas, but when I looked at the complicated, and in some cases extremely verbose, proposals of my colleagues – mainly in Latin – I simply wrote the single word 'Forward'. Tedder, who was our Vice-Chairman almost accused me of levity, but when, in due course, the paper was given to Sir Winston he looked at it, grunted and then growled 'What's wrong with "Forward"?' And so Churchill College got its motto.

The appeal was formally launched on 15 May 1958. As donations were to come to Cambridge I was authorised to send

Ron Purchase round to Barclays Bank in Bene't Street to open an account. According to him he went to the bank and asked if he could see the manager. He was told politely that the manager was busy, and was asked why he wanted to see him. 'I want to open an account,' said Purchase. 'Come, come,' was the reply, 'you don't need the manager for that – one of the tellers will do it for you. What kind of account do you want?' 'Well,' said Purchase, 'it is a little bit complicated; you see I want to open it with a million pounds.'

There was a moment of stunned silence – and Purchase saw the manager.

Looking back at it after some twenty years, I am struck by the smoothness with which the whole operation was carried through. Everyone worked really hard and cooperatively, and I think particular credit goes to Jock Colville whom I consider to have been the mainspring of the whole project. I am proud to have been associated with the project and with him.

I contrived to avoid the hectic days following the public launching of the Churchill College appeal by going off to conduct the honours examinations in chemistry at the University of Malaya in Singapore, and the University of Hong Kong. This was my first visit to South-East Asia, apart from a one day stopover in Singapore on my way to Australia in 1950. I enjoyed it hugely – so much so that, over the years since then, I have taken increasingly to visiting it, and especially that fascinating and beautiful place, Hong Kong. The Malayan part of my 1958 visit was not without excitement. Before leaving London I had been asked by our government to discuss certain matters with the Tunku Abdul Rahman on my way to Singapore. I duly alighted from the intercontinental plane at Kuala Lumpur, stopped the night at a hotel not far from the spectacular railway station of that city, and saw the Tunku the following morning at his residence. After the meeting there was a lunch party during which one of the Malayan military attachés asked me what I was doing after lunch. I told him I was taking a plane to Singapore, whereupon he said he was driving down to Singapore that afternoon, and would be glad of my company if I would like to see something of Malaya. I accepted with alacrity, cancelled my plane reservation, and set off to the south with the attaché who was driving his own car.

A mile or two north of Seremban we were overtaken by a tropical thunderstorm and our windscreen wiper packed up. As more such storms were quite likely at that time of year, we decided to stop at Seremban and have the wiper put right. While the repair was being made, we retired to the local rest-house and had a drink or two. The repair took rather a long time, and when we set out again dusk was fast approaching. As we drove south through alternating jungle and rubber plantations, I noticed that the vegetation had been cut back so as to leave a strip of twenty-five yards or so of open ground on either side of the road. I realised that this had been done to lessen the chance of an ambush by bandits, who had not yet been completely eliminated from the Malayan scene, but thought no more of it. Soon it was quite dark, and, as we drove on and on, I couldn't help noticing that the road was deserted; no peasants with donkeys or goats, such as usually abound, were to be seen. I also became aware that my friend the attaché was getting a bit worried, and was increasing his speed, thereby considerably diminishing our comfort. After what seemed quite a long time we came over the brow of a hill, and there, on the hillside on the opposite side of the valley in front of us, were the flickering lights of a small village or town. 'Thank God!' said my companion, 'Yong Peng at last.' Just at that moment a cluster of lights shot up into the sky above the village, 'Ah,' said I, 'we are just in time for a Chinese wedding or funeral!' 'I'm not so sure about that – they look like Verey lights,' said the attaché, and with that he accelerated and we shot down the hill and up the other side. Before we got to the village there was a burst of machine gun fire in the jungle to our right, but we were unscathed and pulled up in front of a massive wooden gate barring entry to Yong Peng. We made a bit of noise, whereupon the gate opened slightly and a Malay soldier put his head out and asked us our business. Satisfied, he opened the gate sufficiently to let us in then closed and barred it again before taking us to see his commander in a nearby house. The commander, a young officer, told us that we had arrived just as a battle was about to begin, and that we would have to stay in Yong Peng till it was over. Apparently the authorities had learned, through informers, that on this particular night the local bandits planned to attack Yong Peng to replenish their food supplies, and the army had arranged to move in a detachment of soldiers secretly so as to give the bandits a warm welcome. We had arrived just as the first bandits had emerged from a nearby rubber plantation – hence the Verey lights – and firing had begun. The commanding officer was very apologetic, and suggested that we leave our car at the command post and proceed down the main street where we would find a bar at a Chinese hotel called the White House; there, he suggested, we should have a drink and wait until the battle was over.

We took his advice and sat down on the verandah on the street frontage of the White House, which effectively shielded us from the back garden, at the bottom of which a couple of soldiers were ensconced taking occasional pot shots in the direction of the rubber trees. In front of us there was, of all things, a juke-box which two urchins kept feeding with coins, so that the night was filled with the strains of 'See you later, alligator' punctuated, from time to time, with small arms fire from the direction of the back garden. About two hours, and most of a bottle of whisky later, the shooting had ceased, and the commander said we could leave now, although it was clear that he wasn't too happy about it. He offered us an armoured car escort, but my companion would have none of it. If we had an escort (so he said) we would certainly be ambushed, whereas, if we went on our own, the bandits wouldn't bother us. The commanding officer kindly said he would telephone the towns and villages we would have to pass through on our way so that they would expect us and so open the gates to let us through. So off we set again and reached Singapore in the small hours, our journey being without further incident apart from an encounter with a large pig near Johore Baru. There was an amusing sequel to all this when I went along to breakfast at my Singapore hotel next day. In the dining room I met a young

British lieutenant, the son of a friend in England. He was surprised to see me, and asked how long I had been in Malaya. I said I had only arrived in Singapore a few hours before after an encounter with bandits at Yong Peng. At this he expressed great indignation, holding that it was totally unfair that he should have been four months in Malaya looking for bandits without success, and here was I in touch with them within twenty-four hours of my arrival.

Examining in Singapore was quite tame after all this, but it was very interesting. I began to be aware of some of the problems which led to the later separation of Malaysia and Singapore. It was quite striking to see that the student body seemed to be mainly Chinese, with some Indians and very few Malays; in my experience, too, the Malays were the least able students, and gave the impression of having been badly prepared for university education. Hong Kong was, of course, very different in this respect, the student body being entirely Chinese. The University of Hong Kong was largely staffed (although by no means wholly) by expatriates from the United Kingdom; academically its standards were quite high, but it seemed to me at the time to have a quaintly colonial atmosphere. It had a first-class Vice-Chancellor in Sir Lindsay Ride, an Australian trained originally in medicine; he and his wife May were excellent hosts; both had long been resident in Hong Kong, had become part of the local scene, and were highly and deservedly respected throughout the colony. Both Victoria and Kowloon were much less built up than they are today – indeed, as I recall it, the only buildings which might have been called skyscrapers were the headquarters of the Hong Kong and Shanghai Banking Corporation and the Communist Bank of China which stood cheek by jowl not far from the waterfront in Victoria.

By the time I returned to England (after pausing to attend a joint meeting of the British Society for Chemical Industry and its Italian counterpart in Turin) the Churchill College Appeal was well under way and shortly thereafter the Trustees appointed Sir John Cockcroft as Master. Thereupon we appointed a Bursar (General Hamilton) and a Senior Tutor

(J. E. Morrison) to help in organising and planning the new college. Building plans were approved following a competition among architects, and the actual construction was begun about a year later. From this time onwards I kept in touch with progress but was not intimately concerned with the further development of Churchill College.

It was my hope that with less commitment to Churchill College I would have much more time for research; but I had reckoned without knowing that in the autumn of 1957 I was to be awarded the Nobel Prize for chemistry. One of the consequences of receiving such an award is that one has academic honours showered upon one and receives invitations to lecture, open laboratories and so on all over the world; some of these one can decline, but by no means all of them.

So it was that I found myself again visiting India in 1960 where, in addition to lecturing in a number of centres, I received the degree of D.Sc. *honoris causa* from the Muslim University of Aligarh. That really was a party! My wife and I travelled to Aligarh by train from Delhi, and were met and escorted from the station to the university by a troop of cavalry belonging, I presume, to the Officers Training Corps. It was quite an experience to arrive at the huge *pandal* which had been erected for the graduation ceremonies (for several hundred students were taking their degrees at the same function) with an escort of horsemen in highly coloured uniforms and bearing lances with pennants flying. At the ceremony itself I was impressed by the fact that all the girl students wore the traditional black Islamic garb with *yashmak*; I must say that several of them, whom I saw riding bicycles in this dress, looked rather comical – and, I thought, distinctly unsafe!

At home my duties as chairman of the Advisory Council on Scientific Policy and adviser to the Minister for Science still demanded a good deal of attention. I had also, since 1955, been intimately concerned with the International Union of Pure and Applied Chemistry and was a member of the Executive Committee of that body and later, from 1963 to 1965, its President. At that time, the Committee was under some pressure from overseas colleagues to stage some manifestation in the Southern Hemisphere, which had been totally neglected as a meeting location throughout the Union's history. It was finally decided that, with the help of our Australian colleagues, we should break new ground both geographically and subject-wise by holding the Union's first Natural Products Symposium in Australia in August 1960. Our Australian colleagues did a magnificent job and, with full cooperation from State and Commonwealth governments, the symposium, which divided its time between Melbourne, Sydney and Canberra, was a huge success and began a regular series of such symposia under the auspices of IUPAC, which have since been held biennially in many countries around the globe. I had the honour to be President of the 1960 Symposium which was attended by a large number of the leading chemists from every continent. It was a truly memorable event.

My friend Bob Woodward of Harvard was also going to the Australian Symposium and, as he had never previously been in the East, we decided to fly out together and spend a day in Singapore *en route*. I met him on his arrival at Heathrow from Boston in mid-August and we both proceeded to the plane for Singapore; unfortunately we had both forgotten to do anything about cholera injections, so we were hauled off to a medical unit on the perimeter of London airport and forcibly inoculated before being put on our plane for a trip thereby made somewhat less comfortable than it might have been. However, we arrived in Singapore the following evening in time to have a shower and go to a party. Very late, we returned with our host and hostess (Mr Johnson, the Singapore manager of Glaxo Laboratories Ltd, and his wife) to their home for a few hours sleep (before that could be done I remember having to remove an enormous cockroach-like creature from Bob's bathroom, while he stood by petrified, and clearly in some doubt as to whether the creature was real or a figment of a slightly whisky-inspired imagination). The following day we crossed over the causeway and, with our hosts, spent a pleasant day in Johore as guests of the State's chief medical officer (an ebullient Chinese with a fine taste in food) and returned to Singapore to get the evening plane to Sydney and Melbourne.

On our return we were informed that our plane would be at least four hours late, and as a result we found ourselves at another party and were almost poured on to the Australian plane in the small hours of the morning. What with about three nights without sleep, and with sleet falling quite heavily on our arrival at Melbourne, we were not in the best of shape to deal with the mass of reporters waiting at the airport. Bob, however, pulled himself together sufficiently to deliver a strong attack on what he described as the most dangerous characteristic of the Australians – addiction to fresh air and physical exercise both of which he abhorred. The symposium opened in Melbourne then moved after a few days to Canberra and had its final week in Sydney. I had a large black Commonwealth limousine at my disposal in each city, so that Bob Woodward and I drove around in state. The whole trip was punctuated by hilarious incidents in which we were involved but perhaps the most memorable occasion was the delivery of my Presidential Address. I gave it in Canberra in that remarkable hemispherical building which houses the Australian Academy of Science. In the lecture room there, the wall behind the rostrum was ornamented by vertical slats alternately light and dark brown in colour set an inch or two apart. Now, when lecturing without using notes, I have a habit of frequently moving a step or two to right and left while I am speaking. With a background of coloured slats the effect on an audience of such lateral movement is rather like sea-sickness. Since that day in Canberra I have always claimed that, while most of my scientific colleagues both can and do put their audiences to sleep on occasion, few, if any, can emulate me in my feat of making members of my audience actively ill.

The highlight of the Australian trip was the visit paid by a group of symposium participants to the Australian-mandated territory of Papua–New Guinea. This trip was arranged by the Australian government and the Commonwealth Scientific and Industrial Research Organisation (CSIRO) to give a selection of overseas delegates the opportunity to make some assessment of the territory's potential, and to comment, if we wished, on what the Australian authorities were doing there. The group –

about fifteen to twenty in all, including our Australian hosts – was multinational, and Bob Woodward and I were fortunately, if not very surprisingly, included. It was a memorable trip to a fascinating country of jungle, mountain and swamp, with the climate up on the plateau beyond the Owen Stanley Range so beautiful, that I could understand why we found, living in Wau, two retired bankers and their wives from Europe; having set out on a trip round the world, they had found Wau so near to perfection that they simply gave up circumnavigating the globe and settled down there. After spending a couple of days in the Port Moresby area, we flew up in an ex-army DC3 through a high pass in the Owen Stanley Range to Wau – not a large place by world standards, but the headquarters of the local District Commissioner, and the southernmost point reached by the Japanese army in the last war. The grassy airstrip on which we landed was quite markedly sloping; so much so, that, on touching down at the lower end (where, incidentally, the Japanese had been finally halted), the pilot had to switch on power again in order to get our plane up to the top of the runway, where the somewhat rudimentary airport buildings were located. I remember that, while we were thus engaged in struggling up the hill, I observed a group of two or three local inhabitants clad apparently in stripes of blue, white, and red paint, with little or nothing else; they were leaning on spears, and their hideously painted faces were further adorned by what appeared to be bones through their noses, and a collection of rings or discs suspended from the septa thereof. My immediate reaction was to think how thoughtful it had been of the District Commissioner to have a welcoming party specially dressed up to welcome the guests. However, as I drove in the Commissioner's jeep from the airport, I quickly realised that the men on the airport were not specially dressed (or undressed) for the occasion – all the local inhabitants about the place were in a similar state. Subsequently moving around the area and visiting various villages I came to realise that this was the normal attire of these stone-age people. We did see them in all their finery on one day when the Commissioner, having heard that a big tribal feast or sing-sing was being held in one area, persuaded the local chief, whom he knew, to let us come along to the village which had been built specially for the occasion; there we saw the tribal groups from surrounding villages assembling for the feast – all highly painted, with the warriors wearing their headdresses of bird of paradise plumes and performing elaborate formation dances, while the women and children sat around chanting and preparing mountains of food. The New Guinea Highlands can be described as a real man's country. In the Wau district each tribe did one day's work per month on the local roads for the government; on the occasion I saw this day of labour, all the work was being done by the women and children, while the men of the tribe, in their full war-paint, stood around in groups smoking and leaning on their spears.

Wau was really a remarkable place. Some miles up a rough road, following the course of a stream in a nearby mountain valley, there was a clearing in the jungle in which were several buildings or huts, and in which the stream had been diverted into artificial channels, obviously used for gold-washing. Most of the huts were made out of what looked like flattened biscuit tins – a common construction material in that part of Papua – but one was more substantial, being solidly built of wood and stone. In it lived an elderly Scotsman, known locally (and inevitably, I suppose) as Mac, quite alone, save for the thirty or so Kuku-kuku tribesmen and their families who lived in the clearing and were his employees and bodyguard. The women did his domestic chores, while the men washed for gold in the stream. Mac himself told me he had been a gold miner who moved on from one gold rush to another; he moved from Alaska to join the New Guinea gold rush, and, when the latter waned, he decided that perhaps the time had come to retire. What better place was there to do so than the hills about Wau? Here he could have peace with his group of native 'boys', and a pleasant part-time occupation washing for gold in the stream. It seemed to work well enough; I was told by the Commissioner that, each month, Mac came down to Wau to bank his gold, get uproariously drunk for a couple of days, and be carried back up to his home by his Kuku-kuku 'boys' – members, I should add, of the most notorious cannibal tribe in the area. It was said locally, that the proceeds from his gold were sent regularly to his unmarried sister in Dundee, who was getting something like £10 000 p.a. from this source. Mac I found surprisingly well informed, although a month or so behind with the news; he abhorred radio, but was a subscriber to the *Weekly Scotsman* and *The Economist*, which he received regularly, and read from cover to cover. During our visit to him we had a hilarious ceremony (apparently planned in advance, but without my knowledge) at which the Doctorate of Science of the (non-existent) University of Wau was conferred upon me *honoris causa* by the self-styled Chancellor and Vice-Chancellor in the guise of the District Commissioner and his assistant, wearing as robes flowered dressing gowns borrowed from their wives. Mac and his 'boys' formed a suitable arc behind the 'Chancellor' as the 'Senate', while I was made to wear, as academic dress, a very fine beaten bark cape which, I was assured, had belonged to a gentleman who had been jailed a fortnight before for eating his wife.

From Wau we went on through Bulolo and Lae, and saw something of the timber industry and the successful efforts being made to develop coffee and cocoa culture. The whole trip was a very jolly affair, punctuated by many amusing incidents, but also making me, at least, realise the potential of this remarkable country coupled, I confess, with a considerable feeling of unease about the future of the large population of rather child-like stone-age people, who were clearly about to be thrown all too quickly into the maelstrom of the twentieth century. The New Guinea party returned to Sydney, and there dispersed, most members returning to Europe or America, while I went on to pay my first visit to New Zealand, a country in which, like Australia, I had a number of friends – some like Professor 'Bob' Briggs of Auckland (who had been with me on the New Guinea excursion) and J. S. Watt, fellow-students from Oxford days, and others who had been my own pupils in Cambridge. I flew from Sydney to Christchurch, stopped there for a couple of days and gave a lecture, then drove southwards with my former student R. E. Corbett, visiting

Mount Cook on the way to Dunedin and the University of Otago, where he later became Professor of Chemistry. From there I flew directly to Auckland, where I was joined by my wife; together we then drove down through the North Island to Wellington, our final port of call in New Zealand. It being September the weather was cold and rather broken, especially in the South Island, but this did not prevent me from enjoying the great natural beauty of New Zealand. I confess, however, that I was not in any way attracted by it as a place in which to live. The cities and their inhabitants seemed very provincial in their outlook, and I, for one, was conscious all the time of being remote and rather shut off from the world. I have often had this curious feeling of remoteness in the Australian outback, but only in Tasmania and in New Zealand have I experienced it in substantial towns.

My wife and I returned to Australia from Wellington, and proceeded directly from Sydney to Mackay in Queensland with our old friends Charles and Eileen Shoppee; from there we set out on a five-day cruise of the islands lying off the coast, and visited the Barrier Reef itself where, it being low water when we arrived, we were able to land and cruise around in glass-bottomed boats looking at the fantastic gamut of highly coloured plant and animal life that abounds on the reef. The Barrier Reef fully lived up to our expectations, and we were so attracted by the relatively undeveloped Lindeman Island in the Whitsunday Passage that we resolved to return to it. And return we did, for we spent two marvellous holidays there in 1968 and 1974. During the next five or six weeks we visited all the Australian capital cities, spending a most enjoyable time with friends in all of them before returning to Cambridge and normal life again. This was my wife's first visit to Australia; like me, she at once fell in love with the country, and we have been frequent visitors to it ever since.

During the winter and spring of the year 1960–61 I was able to push on with research, especially in the field of the aphid colouring matters, which got more complex and fascinating the farther we explored them. As far as I recall, life was rather uneventful although, outside the laboratory, I had A.C.S.P.

work and made short trips to Germany and Switzerland, for the most part lecturing or dealing with International Union business. In May 1961, however, my wife and I went to the United States, where I received an honorary degree from Yale University. Although I had received a considerable number of honorary degrees from foreign universities, this was the first occasion I had attended (let alone participated in) an American Commencement Day ceremony. It was quite an experience – the procession across the open common in New Haven to the open-air auditorium, where, on a warm early summer morning, the graduations occurred; it was, incidentally, the first time I had seen degrees conferred on graduating students in blocks of fifty or so at a time, rather than individually. To a European like myself, the ceremonial seemed slightly brash, but it was impressive.

After the festivities were over, we were met by my friend Dick Perkin – the creator and President of Perkin Elmer Inc. – whom I had known since my former laboratory colleague, Ralph Gilson, had joined him to develop an English subsidiary. Dick lived in New Canaan, Connecticut, not far from New York, and he had driven up to New Haven to take us back to his home, where he and his wife Gladys entertained us for some days before we went north to Boston and Harvard to visit the Woodwards. Dick Perkin was a remarkable man. Although an extremely successful businessman, he was by no means a typical 'tycoon'. A modest, friendly, well-read person he could be, and was, a delightful host and a loyal friend; his sudden death some years later was a sore blow to all who knew him. When he met us that day at Yale he was driving his latest acquisition – a light grey Bentley touring model built to his specification. He used to relate with glee, how he went down to the quayside in New York to collect it when it was unloaded from the ship which brought it from England. When he got there, the car was standing on the quay, and beside it stood a coloured man who was peering at the controls and, now and then, rather gingerly touching the coachwork. Said Dick, 'Well, what do you think of the car?' 'Dat ain't no motor car, sah! Dat sure am a automobile!' One evening, during our stay with the Perkins on that occasion, we had a barn dance for the children and their friends from the surrounding district. It was a terrific party, and I can still see Dick and my wife demonstrating the Charleston to the youngsters with astonishing vigour! Before returning home after these events, we had our usual spell in Harvard. Bob Woodward was in his usual form, which meant that I found sleep at a premium during my stay; but, as usual, our chemical discussions were stimulating.

Later that summer I attended a meeting of the Bureau and Council of IUPAC in Montreal. I and the other delegates were comfortably lodged in the Queen Elizabeth Hotel, but we had to travel each day a considerable distance across town to our meeting place at the University of Montreal. H. W. Thompson (later Sir Harold Thompson, but then, as now, universally and affectionately known as Tommy) was a member of the British delegation, and he had borrowed a car, either from Perkin Elmer Inc. or from Dick Perkin himself, on a visit to Norwalk, which he made *en route* to the Montreal meeting. On the second morning, Tommy very kindly offered to take my colleague Harry Emeleus and myself from the hotel to the university; we accepted, and set off. To whom any blame should be attached I do not know, but, on a street crossing, we were hit by a car travelling at high speed from our left; our car spun across the street and struck a bank building head on at a fair speed. I was sitting beside Tommy in the front seat; I remember realising that there was going to be a collision, and then the crash into the wall of the bank. I rather naturally assumed I was dead, but after I came to in a few moments, I realised that I wasn't; indeed, apart from a fair amount of blood running down one side of my face from a cut on my head, I seemed more or less all right, and was able to extricate myself and stand up in the street leaning on the car. Tommy had no more than a bruise, but Harry appeared to have broken his collar bone, and possibly dislocated a shoulder. We remained there until an ambulance arrived a few minutes later, and took us to the Queen Victoria Hospital. We were well treated, and after Harry's collar bone had been dealt with, and I had been sewn up, we were taken back to our hotel, each with a bottle of pain-killing tablets, which I fear we put on one side and relied on whisky to restore our rather flagging spirits. We sent Harry back to England, but I carried on, and, when the meeting ended, I went down to New York and spent a couple of days with my son Sandy who, having completed his Oxford Part I in chemistry, was spending the summer vacation working in the research laboratories of Merck Sharp and Dohme in Rahway, through the good offices of my old friend Max Tishler, the company's Director of Research. I thus suffered no harm from this adventure, but it certainly enlivened the IUPAC meeting; I fear the Perkin Elmer car was a write off! The Montreal meeting saw the beginnings of a battle over the position of Taiwan in IUPAC, and in the other constituent bodies of the International Council of Scientific Unions, which was to last for nearly twenty years before a solution acceptable both to Taiwan and the People's Republic of China could be reached. The Taiwanese had inherited the membership of the Republic of China when Chiang Kai Shek and his followers retreated to Taiwan; in Montreal the move to have them expelled was initiated by the Soviet and East European delegates, who wished to see them replaced by the mainland Chinese with whom the Soviet Union at that time had very close relations; it may be more than a coincidence that an amicable solution was only reached following the later breach in Soviet–Chinese friendship.

Meanwhile, at home in Britain matters were stirring in the area of science policy, and government relations with science and industry. In 1959 the office of Minister for Science had been created, and the duties of the Minister described in the following terms:

The Minister for Science is responsible to Parliament for the Council of Scientific and Industrial Research, the Medical Research Council, the Agricultural Research Council and the Nature Conservancy and is Chairman of the five Privy Council Committees to which they report...The Minister for Science will also exercise Ministerial functions under the Atomic Energy Acts and will exercise supervision of the programme of space research...Other Ministers remain responsible for the scientific establishments within their own Departments; but the Minister for Science is responsible for broad questions of scientific policy outside the sphere of defence and is advised by the Advisory Council on Scientific Policy on general questions which relate to the whole field of civil science.

Since the adoption of the Haldane Report at the time of the First World War, responsibility for scientific research had been with the Research Councils and the Department of Scientific and Industrial Research. These bodies, together with the Nature Conservancy, were directly controlled by Privy Council Committees. The Lord President of the Council was, therefore, in effect a Minister for Science, and this was recognised in the reference to his 'responsibilities for the formulation and execution of Government scientific policy' in the terms of reference of the Advisory Council on Scientific Policy, to which I have earlier referred. Taken at its face value, the statement about the appointment of a Minister for Science might have seemed to complicate the problems of science policy, but it did not in practice do so. Lord Hailsham, the then Lord President of the Council, simply assumed the additional title of Minister for Science. Although there appeared thus to be no change in the situation, the additional title had the merit of emphasising the importance of science as a factor in government.

By 1962, however, the feeling was widespread, that Britain was not keeping pace with technological innovation in industry, perhaps because of the very minor rôle we were playing in space exploration which was the glamour subject in those days. As a result, the government set up a Committee of Enquiry into the Organisation of Civil Science under the chairmanship of Sir Burke Trend, Secretary to the Cabinet. This Committee, of which I was a member, reported in 1963. It clarified, and strengthened, the position of the Minister for Science, separating the Research Councils from the Lord President's jurisdiction, and calling for the establishment, in parallel with them, of an Industrial Research and Development Authority, to promote applied research and industrial innovation. Whether the proposals of the Trend Report were wise or not we will never know, for they were overtaken by political events, and were never implemented. In 1963, Government amalgamated responsibility for science and education (including higher education) in a single Department of Education and Science headed by a Secretary of State and with two Ministers of State, one of whom was responsible for science. This in itself was, to my mind, a retrograde step, but it was made much worse when, upon the assumption of power in 1964 by a Labour government with its naïve talk of a 'white-hot technological revolution', the whole set-up was again altered. Science and technology were separated, the former remaining with the Department of Education and Science, and the latter transferred to a new Ministry of Technology with its own Advisory Council. The Advisory Council on Scientific Policy was abolished, and a new, and virtually powerless, body called the Council for Scientific Policy was established. The Council for Scientific Policy was dissolved some years later, and its only real function – to advise upon the apportionment of the science vote among the Research Councils – was transferred to a new body, the Advisory Board for the Research Councils (A.B.R.C.). The Ministry of Technology itself had only a short life, and was soon dissolved, its functions being distributed among several other Departments. The adverse effect of these changes was in a measure offset by the appointment of a Chief Scientific Adviser in the Cabinet Office. The post of Chief Scientific Adviser in which Sir Solly (now Lord) Zuckerman was succeeded by Sir Alan Cottrell, lasted for about ten years, being finally abolished in 1974. The changes made by the Labour government in the mid-sixties were, in my view, a disaster for science in its relation with government, from which we are still trying to recover. The provision of first-class scientific and technological advice – for the two cannot be separated – is essential for the formulation of government policy in an ever-increasing number of national and international issues. In Britain, we began with a highly centralised system for its provision, in the shape of the Advisory Council on Scientific Policy. That Council did not, perhaps, achieve all that had been hoped for, primarily because it lacked the 'back-up' of a good secretariat, and did not always have access to all the information it needed. In my view, however, the decentralisation, and indeed fragmentation, which has followed on the demise of A.C.S.P. has not been successful and a return to something nearer to the old system is badly needed.

The year 1962 was largely given over to chemistry and affairs in Britain, but it also brought an entirely unexpected personal honour in the shape of a Life Peerage, which I accepted. The acceptance was not without its lighter side, for before the barony could be gazetted I had to report at the College of Arms and decide on my title. Since I was well known up and down the world by my family name, and had published all my scientific work under it, I was determined that my title should be simply Todd. Life Peerages were still a novelty in those days, and the College of Arms had not yet got accustomed to people using their own names rather than adopting a territorial designation. The Garter King was clearly anxious to revive Trumpington, which had not been used in a title since the time of the Crusades, and which would be appropriate, since I lived in Trumpington parish; after a struggle I prevailed, and became the Baron Todd of Trumpington in the County of Cambridge. The next problem was my coat of arms; I wanted to use, essentially, the unregistered shield (a chevron with three fox heads) and motto (*Faire sans dire*) used by the Todd family at one time in Scotland. Subject to one modification – the substitution of a 'serpent embowed biting its tail' for one fox head – this was accepted by the Clarenceux King of Arms and registered. The Clarenceux King was anxious to include some symbol for my hobby, but I explained that I had no outstanding hobby which I would wish to include. 'Ah,' he said, 'you are probably one of those men whose hobby is work. Odd, isn't it? I doubt if I have ever done a proper day's work in my life.' The various formalities being over, I was formally introduced into the House of Lords in the summer of 1962 by two old scientific friends, Lord Adrian and Lord Fleck, and took my place on the cross-benches as an independent, bearing no party allegiance. Since I entered the House, there has been a marked growth in the number of Life Peers – and with it a growth in the number of cross-benchers such that they are now a significant force in the work of the House. In recent years there has been much discussion about the future of the

House of Lords. No doubt some reform is necessary, but I am against abolition. I believe that in a parliamentary democracy a second chamber is necessary; the second chamber should have a largely advisory function, and should exert its influence on the lower house through the special qualifications and experience of its members. To do so I believe it should not be an elected body, but rather a body of experts, and not one conducted on party political lines. It seems to me that, however archaic the titles and trappings may seem, a properly operated Life Peerage system provides as good a way as any to achieve this; the hereditary system is indefensible, and should be allowed to lapse.

In the late fifties and early sixties, I had a number of discussions with Dr Albert Wettstein and some other members of the Board of Directors of Ciba Ltd, the Swiss chemical firm, about the promotion of postgraduate student exchanges between Europe and the United Kingdom, to offset the ever growing drift of young scientists from both to the United States for postdoctoral experience; this led to the setting up of the Ciba Fellowship Trust, which successfully developed this idea, and whose action in this respect has been widely copied. Another topic we discussed, which was promoted largely by Wettstein, and put into operation by the Ciba Board, was to tap the undoubted research potential among young Indian chemists, by setting up a research institute under the wing of their Indian subsidiary in Bombay, but with a great deal of freedom as regards research programmes in the pharmaceutical field. By early 1963 the building of the institute at Goregaon near Bombay was complete, Professor Govindachari, an old friend from Madras installed as Director, and all was ready for an official opening by Prime Minister Nehru. This was set for March 1963 and as part of the occasion a special meeting was arranged in Bombay where lectures were to be given by six invited speakers – Vlado Prelog and Alexander von Muralt from Switzerland, Sir George Pickering and myself from the United Kingdom, R. B. Woodward from the United States, and Professor Venkataraman, Director of the National Chemical Laboratory at Poona.

My wife and I decided, with the Woodwards, that the

opportunity for a few extra days' holiday should not be missed, so all four travelled to Egypt, spent a few days in Cairo, and visited Luxor and the tombs in the Valley of the Kings, before meeting Prelog at Cairo airport and flying on to Bombay, where we were royally entertained by the Ciba company. After the opening ceremony at Goregaon, Alison and I, with Bob Woodward and Vlado Prelog, went up to Poona at Venkatara-man's invitation to see the National Chemical Laboratory, of which he was rightly proud, and spend the night with him and his wife in their official residence. The visit was quite hilarious, largely because something went badly wrong at the Poona power station so that all electricity supply failed at frequent intervals. This made Vlado's lecture an uproarious event, the place being plunged in darkness several times while he was speaking. I remember, too, retiring to Bob's quarters for a quiet drink at the end of the evening party, only to have the lights fail after a few minutes. Venkataraman, like the good host he was, was rushing around the guest rooms with candles. What he thought on finding us sitting peacefully having a nightcap in total darkness I don't know, but, in the flickering light of the candle he was carrying, he certainly looked rather sur-prised! We returned safely to Bombay, and wound up the ceremonies there with a big party at a restaurant on the coast a few miles north of Bombay. The highlight was a display of South Indian dancing by Govindachari's daughter Anurada; although only eleven years of age, she was already well known in the world of the dance but remained a rather shy, unspoilt little girl. It was on this occasion, incidentally, that I accepted an invitation from Dr. Käppeli (chairman of Ciba Ltd) to become a Trustee of the Ciba Foundation. From Bombay, the entire party then went off on a beautifully arranged tour to Jaipur, Udaipur, Delhi, Agra and Fatehpur Sikri before return-ing home.

During that same year I made two visits to the United States – firstly to Harvard for Commencement, where I received an honorary degree, and had the interesting experience of visiting a small Liberal Arts College (Colby College in Maine) to see Bob Woodward receive a degree there. Colby was a curious, rather gentlemanly place, evidently patronised by people well provided with the world's goods. I believe there are many such colleges with good academic standards in the United States, although I have only seen one other like it, namely, Widener College, near Philadelphia, where I received a degree much later, in 1976. I have not seen any equivalent institutions in other countries I have visited. My second visit was in October, to attend the celebrations in Washington to mark the Centenary of the National Academy of Sciences, of which I had been a Foreign Associate for a number of years. The Washington celebrations gave me the opportunity to meet, if rather briefly, the young President of the United States, J. F. Kennedy, who was well known to several of my American friends; I found him most impressive, and very different from most of the American politicians I had met. Following the National Academy meeting, I went on to Yale where I delivered, rather belatedly, the Silliman Lectures I had been invited to give some time before. I was well looked after by Joe Fruton, Professor of Biochemistry at Yale, who had doubtless been responsible for my being made a Silliman Lecturer. I ought, however, to feel rather ashamed because, although I delivered the lectures (on nucleic acid chemistry) I never got around to writing them up for publication in book form; I fear I have always avoided tasks like that!

Meanwhile, I had taken on a new job. In January 1963 I was approached by the Vice-Master of Christ's College, who told me that it was the unanimous wish of the Fellows that I be invited to accept the Mastership, which would be vacant in mid-July when the then Master, Professor B. W. Downs, would retire at the statutory age of seventy. At various times before then I had been approached to take vice-chancellorships of various universities and headships of colleges, but had uniformly refused. Now, however, things seemed rather different. For one thing, it was my own college and I knew it and its ways so well that it would not be a great strain to take it over. Again, I had now achieved much of what I could expect to achieve in research, and felt I should now concentrate more on providing the opportunities for younger colleagues to realise their potential. Finally, the Master's Lodge was a fine old house – possibly a bit large, but nevertheless most attractive as a place in which to live. So I accepted; I assumed office on 11 July 1963, but, as fairly extensive modifications had to be done to the Lodge, we did not actually move into it until April 1964. I have always believed that the real reason for this long delay was that the builders' men were keen to keep a nice inside job like this going until the end of winter, remembering the savage cold of the preceding year. I thoroughly enjoyed my fifteen years as Master of Christ's; living with successive generations of undergraduates in college is a most rewarding experience, and leads to the formation of many friendships. I was also very lucky in the key appointments I made. Within a year of my appointment, the college Bursar left to take up a Professorship at the London School of Economics. He, like most college Bursars in Cambridge, was only occupied part-time with his bursarial duties, and, although this was traditional in all but the largest colleges, it seemed to me that a full-time Bursar with no other duties would be desirable, even in a medium-sized college like Christ's. Moreover, it was clear to me that a full-time professional bursar would save the Master a lot of work and trouble. Accordingly, I persuaded the Deputy Treasurer of the University, C. K. Phillips, to resign and become Bursar of Christ's, guaranteeing him terms as good as the university could offer. I was, however, left with another problem. A year or two before I became Master there had been a rather unsavoury row in the college which led to the Senior Tutor giving up his position. His resignation was a serious blow to the college, and I was faced with the problem of working with part-time, reluctant, Senior Tutors with no real policy and no continuity – a disastrous situation for a college to be in. I had always believed that the Senior Tutor should be a full-time servant of the college; the position is too important for it to be in the hands of people who cannot devote their whole time to it. Fortunately I had a great stroke of luck. In 1967, the Commonwealth Fund Trustees would seem to have had a brainstorm, for they let it be known that they were going to cut back or eliminate their Fellowship scheme in Europe and devote their funds to medical research. The Fund had in London a brilliant secretary, S. Gorley Putt, with vast experience in interviewing and selecting young men for Fellowships. Gorley was an old Christ's man and a bachelor, and so had everything I could ask for. I was pretty sure that the Commonwealth Trustees would soon come back to sanity, so, without further ado (or even consulting the College Governing Body), I pounced on Gorley, who to my great joy accepted the post of Senior Tutor in charge of admissions, a position which he held with distinction throughout the remainder of my Mastership. With Phillips and Putt, running Christ's was an easy task, and I believe the College thrived as a result.

My son Sandy completed his Oxford D.Phil. in organic chemistry in 1964 and went to Stanford to do two years of postdoctoral work with Gene van Tamelen; at the same time his contemporary, friend, and by then brother-in-law, Philip Brown (having married my elder daughter Helen) also went as a postdoctoral to Melvin Calvin at nearby Berkeley. Helen, who had just given birth to her first child (actually on my birthday), stayed behind, while Philip went on ahead, and my wife accompanied them to Berkeley in December. There I joined her, after attending a meeting of the IUPAC Executive Committee in Austin (Texas) in December and flying from there to San Francisco. My main recollection of that, my first, visit to Texas, was of a rather aggressively state-proud people, who appeared to tolerate without complaint the most absurd liquor laws I have ever come across. You could buy any liquor you wished in special liquor stores, but apparently could not buy it for consumption on the premises. It was, nevertheless, quite legal to take your own liquor to a restaurant, and there consume it. My first experience of this remarkable arrangement was on Dallas airport; having a bit of time to spare waiting for a connection to Austin, I went into a rather garish pseudo-Hawaiian cocktail lounge, looked at the impressive array of cocktails on the menu, summoned a scantily clad waitress, and ordered one. 'Where is your bottle?' she asked. I confessed that I hadn't got one, whereupon I was informed that she could mix a cocktail for me, but only if I could provide the alcohol! One got used to surprises in a state where the Attorney General was always referred to as General Carr, without his having, as far as I know, any military connection whatever.

My wife and I visited our family in California on several later occasions. The first was on our way to Australia in March 1965, and the second on our return. On that occasion we left Canberra by air for Sydney on the morning of Good Friday, and had hot cross buns on the plane. Our connecting flight from Sydney to San Francisco did not leave until the evening, so we had tea and hot cross buns with the Le Fèvres at their home in Northbridge. Owing to the complication of the date line, we found ourselves having hot cross buns again both *en route* to Hawaii and between Honolulu and San Francisco. The third time was in 1966, when I received an honorary doctorate from the University of California at Berkeley. These visits were particularly interesting, because they coincided with the period of widespread, and at times violent, unrest on the campus. I have mentioned earlier how at Aligarh in 1960 I had a cavalry escort to the degree ceremony; I went one better at Berkeley. The dissident students (abetted, I fear, by some members of the academic staff, who should have had more sense) were particularly incensed because one of my fellow graduands was Senator Goldberg, and they threatened to disrupt the graduation ceremony in the great open amphitheatre on the Berkeley hillside, in which the university ceremonies are staged. The upshot was that I had the distinction of processing to the ceremony with an escort of the Oakland riot police, steel-helmeted and heavily armed. Within the actual arena, no police were visible, but in all the aisles, and around the back and sides of the seated accommodation, were large numbers of huge young men wearing football sweaters who, I suspected, were not there simply out of a love for ceremonial. Partly, I think, because of their presence, when the President, Dr Clark Kerr, in a bold and forthright opening speech suggested that the protesters might at least accord the right of free speech to those with whom they did not agree, and those who would not might leave the auditorium, a small number did, in fact, rise and move sheepishly out. The vast majority, however, remained, and the graduation ceremony proceeded without any further interruption. Looking back at the events in Berkeley in those years, it is hard to understand what all the fuss was about. No doubt the Vietnam war had a lot to do with the feelings of the young people in America, but the dissidents were not all young, and the cults of flower people, hippies and so on, spread all over the world during a period of about ten years – and, indeed, relics of it still persist in the behaviour of some sections of our youth today. I also encountered student unrest in Ghana in 1971, while in 1973 I was invited to accept an honorary degree from the University of Ife in Nigeria, but never in fact received it, since the graduation ceremony arranged for March of that year had to be cancelled because of student riots. For similar reasons, the insignia of my honorary doctorate from the University of Paris had to be delivered to me in London in 1969 by His Excellency the French Ambassador.

In 1966 my son-in-law Philip Brown was awarded a Queen Elizabeth Fellowship to do a couple of years' research in an Australian university. He, with Helen and their daughter Alison, went to Adelaide, and my wife fulfilled her duty as a good grandmother by going out there in the autumn of 1967, to preside at the birth of a grandson, while I went off to Czechoslovakia on IUPAC business. We were both back in Australia in 1968, where I had two duties to perform. First, I was invited as chairman of the Royal Commission on Medical Education, which had just reported, to attend and address the joint meeting of the British and Australian Medical Associations in Sydney in August, and, immediately thereafter, in the capacity of Chancellor of Strathclyde University to attend the conference of the Association of Commonwealth Universities which was also held in Sydney. We were able to spend a fortnight's relaxed holiday on Lindeman Island during our stay, and finished our visit by going to Adelaide for a few days, and then to Western Australia where, at Fremantle, we saw our Adelaide family off on its return voyage to England, where Philip was taking up a lectureship in chemistry at the University of Newcastle; there, in 1970, our second grandson was born.

This accomplished, my wife and I flew from Perth to Johannesburg with an all too brief stop in that enchanting island Mauritius. In Johannesburg I visited the company owned by Fisons Ltd, and had some discussions about the organisation of the fertilizer industry in South Africa. We were only a few days there, but, in addition to seeing the industrial complex at Sasolburg, we visited Pretoria, where I had a talk with the chief of the South African Council for Scientific Research. Johannesburg is not, in my opinion, a very attractive city. Apart from its appearance, which is spoilt by the great heaps of mining waste, I found the general atmosphere depressing, and, in some measure, rather frightening. On the one hand one had the very colonial-style country clubs, and, on the other, the depressing black townships out on the veldt. To see apartheid in action is not pleasant, and I find it hard to understand why the whites in power cannot see the writing on the wall, and move more rapidly to a free society than they are moving at present. Disaster need not overtake South Africa, but even a brief visit convinced me that time is running short.

In the autumn of 1963 Harold Macmillan resigned as Prime Minister, and, after a deal of confusion at the Conservative Party Conference, Lord Home emerged as his successor, resigning his peerage in order to do so. In forming his government he appointed Lord Hailsham (who also disclaimed his peerage) as Secretary of State for Education and Science. The fact that I had just taken on the Mastership of Christ's, and could hardly take leave of absence until I had been there at least a year, enabled me to avoid becoming Minister of State for Science. It was, however, fairly clear that I would be in some danger of renewed political pressure, if the next general election, which was shortly due (and did indeed take place in October 1964), resulted in a Conservative victory. By late summer 1964 I had formed the view that Labour was likely to win the forthcoming election, and I had a fair idea of the kind of reorganisation of science and technology in government which that party, largely under the influence of Patrick Blackett and some other scientific Labour adherents, had in view. Blackett himself came to tell me about their plans, and it was clear to me that I was so much at variance with his ideas that, if Labour were to be returned to power, I would either have to resign as Chairman of A.C.S.P. and be labelled as a man who would only serve a Conservative government, or be sacked for refusing to toe the party line. Neither of these alternatives appealed to me, so I resigned with effect from 30 September 1964, and went off to open the new chemical laboratories at the Technion in Haifa.

David Ginsburg, an old acquaintance, was head of the chemistry department at Haifa, and had been responsible for the planning of the fine new building I was to open on 11 October 1964. He knew my reputation as one who always contrived, on visits abroad, to speak, or at least to be able to make myself understood in, the local language. Rather rashly he had told me a month or two before that this time he would fox me, since the local tongue was Hebrew. This I took as a challenge. I therefore wrote a suitable little speech in English, sent it in confidence to Professor Mazur at Rehovoth (who had worked with me in Cambridge), and told him to translate it into modern, colloquial Hebrew, to write out the translation in accented phonetic script, and to make a tape of it spoken by a native speaker. All this he did, and sent the material to me. It is true that I have always had a certain facility for languages, so I simply committed the whole speech to memory, and reproduced the pronunciation by mimicry from the tape. I thus arrived at the Haifa ceremony well prepared, but having sworn Mazur to secrecy. The ceremony began with a speech from the Minister of Education, followed by others from the Mayor of Haifa and the President of the Technion – all of them in flawless English. I followed with my little speech in Hebrew to an astonished audience which greeted it with great enthusiasm. I must admit that David Ginsburg took it very well, and the whole affair was a great success; the only snag was, that I had great difficulty in persuading the journalists present that I really didn't know any Hebrew at all!

Before returning to England we visited Jerusalem – at that time a divided city. I confess it did not add to the pleasure of sightseeing to be told to keep one's head down near the demarcation line, so as to offer no target to trigger-happy

Jordanian guards on the other side. We also visited the impressive Weizmann Institute at Rehovoth (of which I was to become a Governor a few years later), and one evening, at the home of the British Consul at Tel Aviv, with David Samuel and his wife Reina, we heard on the radio the result of the general election in Britain. The Labour party was returned, although by a smaller majority than most people expected. In one way it was something of a relief to me, since it was widely thought that, had the election gone the other way, I might have been under considerable pressure to take ministerial office. Had such a thing happened, of course, life might have been a little less hectic for, in addition to a lot of overseas commitments, I had acquired added responsibilities at home, which acceptance of ministerial office might have allowed me to shed.

In 1963 agreement was finally reached that the Royal Technical College in Glasgow should have university status. This followed a long, and at times bitter, struggle between 'the Tech.' and the University of Glasgow since the former came into existence as Anderson's University in 1796. Now it was to become the University of Strathclyde, and I was invited to be its first Chancellor. As a born Glaswegian, I was delighted by this honour, although, on appointment, I recalled with some amusement that, many years before, on my first attending a metallurgy class in the old Tech., I had had my overcoat stolen from the cloakroom. My formal installation as Chancellor did not occur until April 1965, when we had a tremendous party in the Kelvin Hall, and I became the first honorary graduate. I have continued to hold the office of Chancellor ever since, and have enjoyed every bit of it. It really has been a joy to be associated with the building up of a modern technological university in which people are not inhibited by the weight of tradition; unlike some of the new universities created in the 1960s, Strathclyde has been a real success, both academically and in its relations with industry. The latter I have watched with particular interest, since, under pressure from my good friend the late Lord Netherthorpe, then chairman of Fisons Ltd, I joined the board of that company as a non-executive director in 1963, and, during the next fifteen years, was in close contact with the actual operation of a large research-based company; I learned much as a result.

Most people in England regard Strathclyde as one of the new universities created as a result of the Report on Higher Education issued in 1963 by a committee under the chairmanship of Lord Robbins. As I have indicated, this is not so; the long drawn-out battle between the 'Glasgow Tech.' and Glasgow University for university status, which would allow it to give its own degrees was finally won before the Robbins Committee reported. The University Grants Committee was, however, aware that the Robbins Report, when it appeared, would propose the creation of a number of other new universities and its chairman, Sir Keith Murray (later Lord Murray) asked Strathclyde to agree that the granting of its Charter should be deferred until a decision had been taken on Lord Robbins' recommendations; the Charter was, in fact, granted in 1964.

The committee set up under Lord Robbins to conduct an enquiry into higher education reported in 1963. I found myself in disagreement with a good deal of the Report, and especially with the proposals greatly to expand the number of universities in the United Kingdom, and to upgrade several colleges of advanced technology by giving them university status. I confess that I was astonished – and still am – by the ill-considered haste with which its recommendations were accepted by the main political parties (largely, I fear, for reasons not unconnected with an approaching general election). I made no secret of my views at the time, and, indeed, criticised the report in the House of Lords when it was adopted, and on numerous later occasions.

The setting up of the Robbins Committee was a response to the growing feeling in the United Kingdom that all was not well with our educational system. We seemed to be educating too few scientists and technologists to satisfy the demands of industry and to make up leeway in the field of industrial innovation; the weakness of our industries in technologically based innovation also encouraged the loss of too many of our ablest scientists and technologists to the United States – the so-called 'brain drain'. On top of all this, the public was frequently provided, through the press and other media of communication, with statistics of the number of young people per thousand attending universities in various countries; these invariably showed Britain to be at, or very near, the bottom of the league. What they did not, of course, show, was the wide variety of institutions listed as 'universities' in different countries; little attention was paid to the pyramidal nature of our educational system, in which the term 'university' was reserved for a small group of institutions designed to complete the education of an élite.

Whatever views one holds about élitism, it seemed to me self-evident that simply to multiply the number of institutions giving education designed for an intellectual élite would offer no solution to our problems. I agreed, of course, with the Robbins view that all who were fitted for university education should have it; but I did not believe that a vast untapped pool of such young people existed, most of them being, supposedly, denied opportunity for advancement for socio-economic reasons. In any case, most of our universities were relatively small, and until each had grown in size to accommodate perhaps ten thousand students I saw no point in creating a rash of small new universities, most, if not all, of which would try to provide the traditional English type of university education, modelled on that of Oxford and Cambridge. The probability that new universities would develop in this way seemed to me the more likely on the basis of past experience. Most of our great civic universities, originally founded as technically oriented institutions, had had their original pattern and aims modified in this way quite early in their development; history, I felt sure, was likely to repeat itself.

An approximate doubling of the number of our universities would, it seemed to me, be not only extremely costly, but, if the number of students was to be vastly increased as Robbins recommended, it would deflect too many of our young people away from the advanced technical and vocational education which was far more needed by the country, and far more suited to the young people themselves. For it must be remembered that, if discoveries are to be made and exploited, far more technicians are needed than scientists and technologists; this did not seem to be recognised in the Robbins Report. If my view that the number of suitable young people who were under existing circumstances denied a university was not as great as many people thought, then doubling the number of universities would almost inevitably lead to a lowering of standards and the presence in the universities of too many students lacking in ability and especially in motivation.

Looking back now over the years, it seems to me that practically everything which I foresaw when the Report appeared has come to pass. I believe, moreover, that the period of student unrest in the late sixties and early seventies was, at least partly, rooted in the Robbins-type expansion which, incidentally, occurred at about the same time in most other developed countries – Germany provides a striking example. One consequence of the creation of a large number of new universities which I did not at the time foresee – although I should have done so – was the way in which the provision of tenured staff for them would mean simultaneous recruitment of a large number of young university teachers from more or less the same age group. This would clearly upset the staff age-distribution in universities, and block the promotion of promising young people coming forward in future years. The economic depression of the seventies has revealed these consequences of our actions all too clearly, and we are now faced with the daunting problem of rethinking some, at least, of our arrangements in the face of financial stringency, which makes our problems even more intractable.

When I resigned the chairmanship of the Advisory Council on Scientific Policy I thought that, apart from occasional participation in the affairs of the House of Lords, I would thereafter be clear of government commitments. In this I was soon to be proved wrong, and, in 1965, I found myself chairman of a Royal Commission on Medical Education. Such a Commission was admittedly long overdue, since there had been no comprehensive study of the subject since Abraham Flexner's report, which was published in the United States in

1925, and the last review of the position in this country was that of the Goodenough Committee in 1942–4; moreover, there was a good deal of unease about the supply of doctors in the United Kingdom, and the ever-growing reliance on immigrant doctors to keep our health services going. I confess that I was at first surprised at being asked to undertake this task; further consideration, however, led me to the view that I had qualifications which made me a rather obvious candidate. For one thing, not being a member of the medical profession, I had no axe to grind; I was also experienced in work with government and government departments, and, in my scientific career, I had always had close contact with the medical sciences and academic medicine. In addition, of course, I had, through Sir Henry Dale and the Nuffield Foundation, many other contacts with the world of medicine, and I was fortunate in having a first-class body of members of my Commission – able, imaginative and hard-working men and women – to all of whom I am deeply grateful, and whose continuing friendship I cherish.

The report of the Interdepartmental Committee on Medical Schools (the Goodenough Committee) reflected the growing dissatisfaction with many features of medical education at that time; the changes made as a result of it were smaller than had been hoped, partly because of the incidental effects of the National Health Service Acts. The institution of the National Health Service must rank as one of the greatest social advances in our history but, for a variety of reasons, its institution was bound to involve acceptance, at least for a time, of the main features of existing medical education and the structure of the profession, not all of which were desirable. The Commission was, therefore, faced with a formidable task which we completed in about two and a half years.

In the course of our studies we took evidence from around 400 individuals and organisations and visited medical schools in the United Kingdom and in a number of other countries; we also examined the various systems of medical care currently in operation, from the highly centralised polyclinic systems of the Soviet Union and other East European countries, to the

Kaiser Health Plan in California, and the rather complex hospital system in Japan. Our studies had their lighter moments. On questioning representatives of the Physiological and Anatomical Societies as to the amount of time which should be devoted to their respective disciplines in an undergraduate medical course, the only answer I could elicit was that each of them wanted no less time than the other. Again it was interesting to be told that the chairman of the General Medical Council in the United Kingdom and the French Minister of Health were both slightly worried about the possibility of Britain joining the European Economic Community. The one felt that Britain might be flooded with ill-trained doctors from France, and the other that France might be similarly flooded with ill-trained doctors from England.

It was clear to me from the outset that recommendations made by the Commission would be of little value unless they provided an adequate answer to the country's need, not just in the immediate future, but for at least a couple of generations. We therefore had to develop a picture of a likely pattern of medical care in the future, and to use it in forecasting both manpower needs and the general pattern of medical education, which was sorely in need of review. The education of a doctor, originally little more than an apprenticeship in which the aspiring student 'walked the wards' with leading practitioners, has retained more of this character than most other professions. The Medical Act of 1858, which still largely dominated medical education when the Commission began its work, was drafted on the assumption that at the end of a few years' undergraduate training, the emergent doctor would be sufficiently competent in medicine, surgery and midwifery, to set himself up in general practice. This assumption, long since totally unrealistic, depended on the view that the essential object of the undergraduate medical course was to produce a safe and competent general practitioner. Not until 1967 did the General Medical Council make any substantial changes in this system of education, which, over the years, had militated against the institution of postgraduate training, and had consolidated the division between the consultant physician or surgeon practising in the major hospitals, and the general practitioner who was without access to beds, and professionally considered to be of inferior status. We clearly could not put matters right at a stroke, but, in our report issued in the spring of 1968, we set out a series of proposed changes (largely a distillate of the views we had received from those who gave evidence) which would, in due course, lead to the desired end. This is hardly the place to discuss these in detail, but, basically, we recommended reorganisation and broadening of the under-graduate curriculum, the introduction of organised post-graduate professional training for all specialities (including general practice) together with some changes in the existing career structure and the introduction of vocational registration.

We devoted a section of our report to the problems of medical education in London with its twelve medical schools, and their associated teaching hospitals, for the most part located within a small area of dwindling population. Our proposals involved reducing the number of schools to six by a process of twinning, and the association of each of the new schools with a multifaculty institution. These proposals were violently opposed by the medical traditionalists, and by the then Principal of the University of London, and were therefore in effect rejected, although, with the passage of time, some parts of our ideas have been put into effect (e.g. in the case of St Bartholomew's Hospital, the London Hospital and Queen Mary College) and I have no doubt that in time more mergers will occur. I found it difficult not to be amused when, in 1979–80, the situation in London had become so manifestly unsatis-factory (as I knew it would if our proposals were ignored) that a new committee under Lord Flowers was set up by the University of London to examine it afresh. The solution proposed by Lord Flowers and his colleagues differed little in general principle from ours, save that they laid no stress on association with multifaculty institutions; like us, they pro-posed the pairing and amalgamation of the London medical schools, but they also recommended the closure of one of them (the Westminster), a recommendation bound to cause a storm.

My own belief is that the Todd Report, had it been acted upon promptly, would have given a better long-term solution, but, however that may be, the Flowers Report has been attacked by the medical traditionalists just as ours was, even after the lapse of ten years during which conditions had further deteriorated. The failure to follow the recommendations of Todd or Flowers, or a distillate of both, will merely postpone and render even more difficult the inevitable reorganisation of medical education in London. It is amazing how ostrich-like our medical colleagues can be!

It was not to be expected that the report of a Commission such as ours would be 'accepted' by government, for much of it dealt with matters over which government could have no control. Government did, however, accept our recommendation that new clinical schools should be instituted in the Universities of Cambridge, Southampton and Leicester, all of which are now in being, giving us the capacity more nearly to meet the growth of demand for doctors in the future. Despite suggestions to the contrary in recent public statements, I still believe that our estimates of demand will be in due course vindicated. Recent changes in the organisation of the National Health Service are also moving in the direction we suggested. In other areas, too, I am not dissatisfied with the progress made since our Report was published; there have been substantial reforms in undergraduate teaching courses, and postgraduate training has undergone much development. It is true that our views on the career structure in medicine have not yet found favour, but it is my own view that they will be adopted in due time; professions as long established and as conservative as medicine are invariably slow to move.

Broadly speaking, the so-called Todd Report was well received, and had a considerable impact in the United Kingdom. Perhaps more surprisingly, it aroused considerable interest not only in Commonwealth countries in which medical education is mainly along British lines, but also in the United States where the pattern is different. The degree to which any of our ideas have been adopted there (if at all) is unknown to me, but I recall the great interest shown by professors and deans from many of the American medical schools when I presented our findings to them at a large meeting of their Association in Houston in November 1968. I found that there was much more on which we could agree than I had expected. But perhaps I should not have been surprised; medical education and medical practice are under the same pressures everywhere. Every new discovery in medicine creates a demand for medical care which did not exist before. This is a situation where invention becomes the mother of necessity, and the demand for more and more health care is insatiable. Under these circumstances all countries will inevitably be driven along the same paths in their efforts to deal with it, and the patterns of medical education will become more and more uniform as a result.

During the 1960s I did a great deal of overseas travel, visiting and lecturing in many different countries. It was during this period that I first visited Africa. In 1965 I was invited to visit Accra where my former pupil, J. A. Quartey, had become Professor of Chemistry in the University of Ghana. I was guest of the recently formed Ghana Academy of Science, and my visit was timed to coincide with that of Queen Elizabeth and the Duke of Edinburgh, so that I saw Accra very much *en fête*, living, as I did, downtown in the Ambassador Hotel. Admittedly, the festive spirit was not very evident among the rather silent and gloomy Russian technicians who seemed to make up most of the company there, but we had flags, fireworks, etc. Kwame Nkrumah was still in power, and I met him once or twice; I was not over-impressed for, although he seemed to me to be an astute politician and something of a demagogue, he had a rather low intellectual ceiling. As a result, he was a ready prey to the rather shady bunch of sycophantic advisers who were around him. He certainly had some foolish ideas. On one evening while I was there, a great banquet was arranged (for men only as I recall it) at which the Duke of Edinburgh was to be formally invited to be Patron of the Ghana Academy of Science, and I was to be admitted as an Honorary Member. On the morning of the day appointed for the ceremony I was breakfasting on the terrace at the Ambassador, when Joe Quartey arrived, accompanied by Professor Ernest Boateng who was, I think, Secretary of the Academy at the time, and who appeared very upset. He told me he had just come from Nkrumah, who had told him that he had decided that the Academy should become a much grander body, and that he should lead the way; accordingly he, Nkrumah, would announce at the banquet the creation of the Pan-African Academy of Science with himself as President, and ask the Duke to be Patron. Boateng, like me, thought that to do such a thing without prior international consultation would infuriate the other African countries, and that the Duke would certainly have to decline the invitation to everyone's embarrassment. We decided we had to take some action, but found that Nkrumah had already set off with the Queen to open the new Volta Dam; the matter had therefore to wait until the evening. Fortunately, between Nkrumah's return to Accra and the banquet, we got him to change his mind and the banquet went ahead as planned; no Pan-African Academy was mentioned, but it was a near thing!

I greatly enjoyed that first visit to Ghana, despite the undertone of discontent with the regime that I found among the academics and other educated people. The people seemed gay and friendly, and the enormous market-mammies in the Accra market had to be seen to be believed. Apparently, the local inhabitants admire the physical proportions of these ladies as much as their business abilities, and find European women rather feeble looking. At the time I speak of, my friend Joe Quartey had a houseboy (actually a middle-aged man) who looked after Joe and his wife Patience in their house on the edge of the university campus at Legon, which is, I suppose, three or four miles from Accra. The houseboy, who had never been as far as Accra in his life, decided he would like to see the Queen, and so he set off early in the morning and walked to Accra to see the Queen perform some ceremony. He duly returned in the evening and resumed his duties without making any comment on his day's outing. The following morning Joe asked him if he had indeed seen the Queen. 'Yes,' came the monosyllabic answer. 'But is that all you have to say about it?' said Joe. 'Huh! She no be big strong woman – no be proper queen!' Parts of Ghana are beautiful – I remember the great rollers coming in on the glittering white sands at Winneba, and the dense forests on the road up to Kumasi. While I was there, Nkrumah was busily building up the newly created Technical University at Kumasi, incorporating in it what had been a first-class technical and agricultural school (which I felt would have been more useful than the grandiose institution being developed in its place). At the time of my visit there was a bit of a row going on between the university administration and a student residence on the outer rim of the campus; in accordance with current happenings elsewhere, I was not surprised to hear that the residence was locally called Katanga and its warden Tshombe! Kumasi was an interesting place – quite different from Accra. Tribal feeling seemed strong, and the Asantahene was held in great respect as effective king of the Ashanti people.

My wife and I visited Ghana on two later occasions, first during the military regime of General Ankara, and in 1971 in the period of civil government prior to the takeover by General Acheampong, when I gave the Aggrey–Fraser–Guggisberg Lectures. On both visits we stayed in the university's guest bungalow high up on the hill behind the main university buildings with a fine view over Accra towards the sea. During these visits the university was having student troubles, rather like its sister institutions in the U.K.; as usual, the basis of the protests and demonstrations was never very clear and, as far as I could see, there was no violence. I spent a couple of hours one day in the Union with the student leaders who had been complaining that they were never allowed to meet distinguished visitors. At the end of our talk, I asked what they were doing next, and they said they had an appointment with the Vice-Chancellor, Dr Alex Kwapong, in his office up the hill at the far end of the campus. As I was only going across the road to the chemistry department and would be there for the next hour or so, I suggested that, it being a hot day, they should all pack into my large black official limousine and be driven up to see the Vice-Chancellor. This suggestion was greeted with enthusiasm, and off they went. What I did not learn until that evening was that the students had arranged the meeting with the Vice-Chancellor to complain that he provided no transport for them from the gateway by the bus stop on the main road from Accra to take them up the half-mile avenue leading to the central university buildings where the main lecture rooms were! It is not, however, true (although it has been said) that, on the following day, a student demonstration was held accompanied by the chant 'Kwapong out, Todd in!' The Ghanaians are a friendly, hospitable and able people living in a country with great potential; it is sad to see them plagued by political instability and consequent economic disruption. One can only hope for better things in the future.

On most of the trips made in the sixties to chemical conferences, Bob Woodward and I travelled together and our friendship became even closer. Indeed, there were times when our appearance together at symposia made us feel rather like a travelling vaudeville act. But the highlight of the decade was when Bob (in my view belatedly) was awarded the Nobel Prize for chemistry in 1965, and I was able to see him receive the prize in Stockholm.

The 1970 Indian Science Congress was held at Kharagpur near Calcutta, and I was invited to be present at the Opening Ceremony performed by the Prime Minister, Mrs Gandhi, on 3 January, and to deliver a plenary lecture. By the time January was approaching I found myself with other commitments added to this – the dedication of the new laboratories of the National Drug Research Institute at Lucknow, and a mass of lectures and other engagements in Delhi and Bombay. My wife and I left London by air on 1 January but, on arriving over Dum-Dum airport at Calcutta the following morning, fog was so thick that the aircraft, despite cruising around for an hour or so, could not land. As fuel was beginning to run low, the pilot decided he must carry on. The end result was that we were deposited in Hong Kong, had some four hours in a hotel there, and flew back to Calcutta arriving in the small hours of 3 January. After a further two or three hours we set off again by road to Kharagpur, which we reached just in time for the opening ceremony. It is fortunate that I was seated behind Mrs

Gandhi, for I am afraid I slept through most of her address, and indeed I was pretty whacked by the time I got through the formal lunch and delivered my own lecture. On the following day I had to attend the Congress meetings until we were taken off to Calcutta again to get a plane to Lucknow. As luck would have it, engine trouble delayed the plane and we arrived late at night in Lucknow only to be on parade again at 9 a.m. the next morning to dedicate the new laboratories of the Central Drug Research Institute, make a speech, have lunch, listen to the young men giving me an account of their research, and board the plane for Delhi. The pace was much the same in Delhi, and after that in Bombay where we spent about a week. I remember getting close to the point of collapse after about two days there; I spent an afternoon in my hotel room feeling very unwell and more exhausted than I have ever felt before or since; but I recovered, and completed my engagements according to plan.

At the time, I put down my feelings in Bombay simply to exhaustion, but now I am not so sure. Early in March 1970 I was struck down by a severe heart attack in Cambridge. Fortunately I was taken at once to Addenbrooke's Hospital and really well looked after by my cardiologist, Dr H. A. Fleming, and the hospital staff, all of whom I knew well since I had been Chairman of Governors of the United Cambridge Hospitals since 1968. After emerging from intensive care, I was told I must stay in hospital and remain effectively *incommunicado* for about eight weeks. My diary was fortunately relatively empty at that time. My next major engagement was the Commencement Address at the University of Michigan and that could be, and was, easily postponed until 1971. Thereafter I was due at Durham in September 1970 as President of the British Association for the Advancement of Science, but I reckoned I would be fit in plenty of time to fulfil that one. The big problem appeared to be how to pass the time, since I was told quite firmly that I must on no account work. I didn't see how I was going to survive on detective novels and the radio, so I cudgelled my brains for an answer. I found it partly as a result of some comments by my friend C. P. (Lord) Snow – I decided to learn

Chinese as a means of keeping my mind occupied. So I sent down to Heffers bookshop for such books as they had on the Mandarin dialect, and sent word to Henry Chan, a Chinese research student in my laboratory, to come along to see me to discuss the Chinese language. I am afraid it caused him quite a bit of concern, for the message was delivered to him 'out of the blue' on 1 April; but he did come along, helped me select the best texts, and I got under way. I found the exercise very rewarding. Chinese proved a very interesting language and, even if I did not come out of hospital a Mandarin scholar, I had learnt the rudiments sufficiently to enable me to build on them subsequently when time permitted. I fear, however, that I was too old to have any real hope of learning the 2000 or so characters necessary for reading even a newspaper. Still, with the increasing use of pinyin transliteration in modern China, the situation for the foreigner is slowly improving.

When I emerged from hospital and was fully fit again I decided that it would be wise to cut down on my commitments and take things a little more easily. I began by shedding a number of committees, and cut down severely on public lecture commitments. It also seemed that I now had the chance to put into practice what I had often told my family I would do once I passed the age of sixty, *viz.* retire from the headship of the University Chemical Laboratory. I had, during my professional life, seen great departments such as Cambridge, Manchester and Oxford badly scarred and well-nigh destroyed by having professors who stayed on after they had ceased to be effective, and I had always vowed I would not make that mistake in the school I had built up in Cambridge. I had indeed already made a move by importing, only nine months before my illness, a young and brilliant professor in A. R. Battersby (who had indeed been an undergraduate member of my school in Manchester during the war period). The way was thus clear for me to show that I could practise what I preached, and I gave notice that I would resign my chair and take Emeritus status on 30 September 1971. I fulfilled my obligation as President of the British Association for the Advancement of Science in 1970, and delivered a fairly hard-hitting Presidential Address entitled 'A Time to Think' in Durham Cathedral (the only place in Durham large enough to accommodate the audience). The address, in which I set out at some length my criticism of the post-Robbins expansion for our universities, did not please the political left; it was leaked beforehand, and the gaiety of the occasion was increased by a demonstration in the Cathedral Yard by a group of left-wing adherents of a body called (I think) the 'Society for Social Responsibility in Science', all attired in white sheets and affecting to be victims of chemical and biological warfare. 'A Time to Think' was published in *Advancement of Science* 1970, **27**, 70; on re-reading it after ten years I find little in it that is not equally relevant today. For this reason, and because it sets out my views on a number of topics of wide interest and importance, I have decided to reproduce it in full as Appendix I (p. 205) to these memoirs.

I kept my word and formally resigned my Cambridge chair in 1971. To mark my retirement the University Chemical Laboratory held a large dinner – so large, indeed, was the guest list that it had to be held in the Hall of King's College. I was deeply moved to find included among the guests not only former colleagues and students from the Manchester and Cambridge days, but also old friends from my student days in Glasgow, Frankfurt, and Oxford, as well as Sir Robert Robinson and R. B. Woodward. It was a tremendous party, and one notable consequence of it was the creation of the Toddlers' Club, an exclusive dining club of eighteen members which I have already described (p. 71).

Even before I resigned my chair it was clear that my life would not be very much easier as a result. Following the death of Sir Frank Lee in the spring of 1971 I came under heavy pressure from the Vice-Chancellor and the University Treasurer to succeed Sir Frank as chairman of the Syndics of Cambridge University Press. They were most insistent and, particularly as I had refused to allow myself to be nominated for the Vice-Chancellorship, I felt I had to accept.

My refusal to stand as a candidate for the Vice-Chancellorship – the highest office in the University of Cambridge – perhaps calls for some explanation. My reasons were simple enough. In the University of Cambridge the Vice-Chancellor holds office for only two years, the running of the university being largely in the hands of its permanent officials. In my opinion, two years is too short a period in which to initiate and put into operation any reforms which experience might show to be desirable. In this sense, the Vice-Chancellor seemed to me to be largely a figurehead with little real power, but carrying a great deal of responsibility if things went wrong. Responsibility without power has never appealed to me, and I believe that the Vice-Chancellorship of the university should be held for a longer period (as in Oxford) or be a permanent appointment as in all other universities in the United Kingdom.

Already on my first visit as chairman to the offices of the University Press I was horrified; the Press was to all intents and purposes bankrupt, with a soaring overdraft and with sales and receipts dwindling, so that they were unable to cope with rising costs. I believe that only the knowledge that behind the Press stood the university with its great resources had kept the bank from calling a halt. Fortunately, even I could see that the Press could be made to function profitably and that all that was wrong with it was bad management. The Press was managed by the Syndics, a group of academics appointed by the university on a system of rotation; no doubt they were excellent choosers of scholarly books, but they evidently thought that the Press should be run just like a university department; they also seemed wedded to the idea that Press staff should not only be, as far as possible, academics themselves but also be remunerated on university scales without reference to the rates paid in the world of commercial publishing. Fortunately there were some members of the Press who realised the position and were looking for a lead which could only come from the top. I had a few busy months at the start but was lucky enough to get things on the right lines quickly. In these initial moves I received great help and encouragement from R. W. (Dick) David, the University Publisher, who was well aware of the problems and who, quite unselfishly, sought with me to reorganise the Press, even if it detracted from his own position of authority; I shall always be grateful to him for his help. It was evident that the first essential was to appoint a really first-class managing director/chief executive with experience in commercial publishing. Since it was obvious that to get such a person one would need to pay the going rate in competition with commercial publishers, I decided that the best thing to do would be to engage the right man and only tell the Syndics about it after the deed had been done. In this way I was able to appoint a really brilliant executive in Geoffrey Cass, who had previously been associated with Allen and Unwin Ltd. He took office with us on 1 January 1972, and from that day the Press never looked back. Within a year it was back on the rails, and by the time I had to resign (with great regret) my chairmanship at the end of 1975 when I became President of the Royal Society, we were not only making very substantial surpluses but had built up an extremely strong cash and assets position. Although the success was undoubtedly due to Geoffrey Cass, I like to think that I played some part in what was an important rescue operation for the University and for academic publishing.

In March 1973 Sir Geoffrey Gibbs resigned as chairman of the Managing Trustees of the Nuffield Foundation and I was appointed in his place. As the Foundation had been in existence for more than twenty-five years and had an efficient administrative organisation at its London headquarters in Regents Park, the duties of its chairman were not unduly onerous. It seemed to me that we had rather neglected our advisory committees in Australia and New Zealand and that it would be a good thing for the chairman to visit them and discuss programmes and policy on the spot. It happened that I had been invited to deliver the Centenary Oration at the University of Adelaide in 1974, so that it was possible to fit in the Nuffield visits on the same trip. Accordingly, in August 1974, my wife and I travelled to Adelaide and, after the university's centenary celebrations, we visited various centres in Australia on Nuffield matters, and spent a memorable fortnight on Lindeman Island before going on to New Zealand for our second visit to that beautiful country.

We flew from London to Adelaide *via* Mexico, where we

stopped off for an all too short visit. In addition to seeing the marvellous collection of antiquities in the great archaeological museum in Mexico City, we were only able to visit two pre-Columbian temple complexes at Teotihuacan and Tula, but that was enough to make me resolve to return to that beautiful and fascinating country at the first opportunity. Prior to our visit I had been interested in the history of pre-Columbian America, and I knew a fair amount about the Toltec and Aztec civilisations; but I was quite unprepared for the breathtaking size and beauty of the Temples of the Sun and Moon and the huge ceremonial avenue of the Teotihuacan complex. Teotihuacan will always rank in my mind with the great temple at Karnak in Egypt, which is about equally majestic and awe-inspiring.

Adelaide University celebrated its centenary in August 1974 during a spell of warm sunny weather which ensured that all the ceremonial and pageantry went off without a hitch. We met many old friends there, including, of course, the Vice-Chancellor, Geoffrey Badger, and his wife, and we were particularly pleased to meet again Sir Mark Oliphant who, aided by his wife, was doing an excellent job as Governor of South Australia and endearing himself to the ordinary people of the state by his friendliness and informality. I always find Adelaide with its wide streets and balconied houses a most attractive city. It has about it a rather quiet, genteel, air which is also noticeable in the Adelaide Club, of which I had the honour to be a member during my stay. The club preserves much of the old-fashioned courtesy and formality now fast disappearing from many of the London clubs on which the Australian clubs were clearly modelled. I have been a temporary member of several others – the Weld Club in Perth, the Union Club in Sydney and the Melbourne Club – and have found them also to be, if anything, more English than their London counterparts.

Lindeman Island, where we went with our friends Lloyd and Marion Rees after the Adelaide ceremonies, was, as on our earlier visits, a really magnificent place for a relaxed holiday in the sun. The beautiful beaches, the informality of the single hotel, and the eerie stillness of the bush in the centre of the island, broken only by the occasional screech of a currawong – all these were much as before. But there was, I thought, an ominous portent of things to come in the admittedly rough and ready nine-hole golf course, which had been carved out of the bush near the island's only landing strip. I fear that this may be the prelude to the kind of so-called development which has already converted a number of other islands between the Queensland coast and the Great Barrier Reef into raucous holiday resorts. I can only hope that I am wrong!

On our New Zealand visit we made an extensive tour of the South Island with Sir Malcolm Burns, chairman of the Nuffield New Zealand Committee, and his wife as our hosts and companions throughout. The weather was cold and not infrequently wet as we went around – Mount Cook, Queenstown, Te Anu, Arrowtown, Manipouri and Milford Sound – but it was all most enjoyable. I found the rain forest in the far south-west quite fascinating; in a land where the annual rainfall averages well over 300 inches, the forests are dominated by lichens of every colour. New Zealand is a country of great scenic beauty; it is a pity that its towns and cities are not equally inspiring.

Two years after I assumed the chairmanship of the Nuffield Foundation I was – much to my surprise – nominated for the Presidency of the Royal Society, which office I assumed on 1 December 1975. To be elected President of the Royal Society is the supreme accolade in British science and in my case it had an added sentimental value. My father-in-law, Sir Henry Dale, had been its President from 1940 to 1945 and did indeed formally admit me to the Fellowship in 1942; he, too, had been awarded a Nobel Prize and was a member both of the restricted German Order Pour le Mérite and of the British Order of Merit. I had myself been made a member of Pour le Mérite in 1965 and when, in 1977, I had the honour to be admitted to the Order of Merit it put my wife in a unique position!

8

The Royal Society

—

The Royal Society or – to give it its full title – the Royal Society of London for Improving Natural Knowledge, is the oldest of the existing national scientific academies and it enjoys an immense prestige in the world of science. Its origins go back to about 1645 when a group of scholars made a habit of meeting together in London and, during the Protectorate, partly there and partly in Oxford, but it was formally founded in 1660 and received its first Royal Charter in 1662. Unlike many national academies it confines its activities to natural science, both pure and applied. One of its earliest secretaries, Robert Hooke, in 1663 laid down rules for the Royal Society as follows:

The business and design of the Royal Society is – To improve the knowledge of naturall things, and all useful Arts, Manufactures, Mechanick practises, Engynes and Inventions by Experiments – (not meddling with Divinity, Metaphysics, Moralls, Politicks, Grammar, Rhetorick, or Logick).

These rules are still followed, and define fairly accurately the scope of the Society's activities. Over the three centuries of its existence it has had frequent contacts with British governments, and has given advice and assistance in matters of policy – for example, in the setting up of the Royal Observatory at Greenwich and (much more recently) of the National Physical Laboratory; but it has always kept clear of political involvement.

Like all human institutions, the Society has had its ups and downs. Periods in which it was vigorous and influential have alternated with others when it was relatively inactive and inward looking; it has, however, always been steadfast in its recognition and support of outstanding ability in science. When I became President, the Society was emerging from a rather difficult period. In the 1930s it had been on the whole quiescent, and, indeed, had something of the air of a scientific gentlemen's club; the war of 1939–45, however, saw it plunged into the problems of scientific policy as the allied governments sought (with success) to harness science to the war effort. By the end of the war, science and its potential stood high in political esteem, and the Royal Society was supremely well placed to fill the role of scientific adviser to government in developing the new post-war world. Unfortunately that chance was missed. Dale, who had been President during most of the war, went out of office in 1945 and was succeeded first by Sir Robert Robinson and then by Lord Adrian, neither of whom had any interest in political or governmental matters. Adrian's successor, Sir Cyril Hinshelwood, was again essentially a scholar and a scientist and occupied himself largely with the celebration of the Society's Tercentenary and with its foreign relations, while Sir Howard (later Lord) Florey in his turn was fully occupied with the problem of reorganising and rehousing the Society in new premises in Carlton House Terrace – a task which involved, among other things, much detailed negotiation and extensive fund-raising. When Patrick (later Lord) Blackett took over in 1965 he certainly moved to increase the Society's rôle in national and international affairs, but, by aligning himself and also endeavouring to align the Society with the political party then in power, he broke one of Hooke's rules and, in my view, damaged the Society in its external relations. Under my immediate predecessor, Sir Alan Hodgkin, repair of that damage and recovery was put in train, but still had some way to go. I decided that I should endeavour (1) to increase the influence of the Society in providing government with advice on scientific aspects of policy while remaining totally independent; (2) to increase to the maximum extent its support of research in the then current climate of financial restrictions by developing and extending its system of research professorships and fellowships; (3) to develop closer relations with applied science and engineering and (4) to strengthen its international relations. The degree to which these efforts were successful is for others to judge, but they are, in any case, too recent to permit a fair judgement.

One of the duties of the President of the Royal Society is to deliver each year an Anniversary Address to the Society at its annual meeting on St Andrew's Day. Although it had not been customary for Presidents to devote their Addresses to current problems, I decided that my Anniversary Addresses should be a vehicle for my views on matters of public concern. This I endeavoured to do, and in each of my five Addresses I dealt with matters related to science which were of current concern to the Society and to the public at large. Because of their importance as indicators of my views, and because quotations from them would hardly do justice to these views, I have included the essential parts of them as appendices.

During my first year of office I found myself much involved with problems relating to the freedom of science and of scientific enquiry, and with much public agitation about the treatment of certain scientists in the Soviet Union and in some South American countries. My views on these matters are set out in my Address delivered on 30 November 1976, of which I append the relevant portion as Appendix II. The end of 1976 brought with it other problems. The President of the Royal Society relies heavily on the Executive Secretary and four Honorary Officers – the Physical, Biological, and Foreign Secretaries and the Treasurer. Although having had only one year in office, I was faced with the need to appoint three new Officers – a Treasurer, a Biological and a Foreign Secretary – the holders of these posts being due to retire at the 1976 annual meeting. I was fortunate enough to make three excellent appointments, but something approaching disaster struck only some three weeks after the Annual Meeting. Sir David Martin, who had been Executive Secretary for some thirty years, died suddenly and without any prior warning as a result of a massive heart attack shortly before Christmas 1976 – only a couple of hours after he and I had attended the annual staff Christmas party. David was not just a dear friend of many years' standing, but he really was the linchpin of the Society and his tragic loss put me in real trouble. Things might, however, have been worse; at least I had appointed three outstanding men as Honorary Officers – John Mason, David Phillips and Michael Stoker – and Sir Harrie Massey was continuing as Physical Secretary. Harrie was a real tower of strength, and in Ronald Keay, David Martin's deputy, we were lucky enough to have a man of experience and ability who could take over the duties of Executive Secretary. So we survived, and were able to carry on with the programmes I had in mind.

In January 1977, in pursuit of my wish to strengthen our overseas relations, I visited the Venezuelan and Brazilian Academies of Science, delivering the Isaac Newton Lecture to the former in Caracas. This was my first visit to South America and, although Venezuela and Brazil are very different from one another, both of them conveyed to me the same uneasy feeling that I have in India, where similar extremes of wealth and poverty exist side by side. But there was no mistaking the warmth of the welcome my wife and I received; I am sure that our visit was well worth while and not only helped materially to strengthen relations between the Royal Society and these South American Academies, but in a wider sphere between these two countries and the United Kingdom.

During 1977 the recession was beginning to bite into research funding, and there was much unease among the general public and among politicians about the 'relevance' of university research and about whether it should be oriented directly to industrial needs; among academics, too, the combined effects of the post-Robbins expansion and financial stringency were widely discussed. My views on these matters formed the basis of my 1977 Anniversary Address and are reproduced in Appendix III, while in 1978 (see Appendix IV) I rejected the gloomy picture of the future painted in the Club of Rome's *Limits to Growth*, and returned again to the problem of freedom of science and the need to beware of external control and direction. Discussion on the right of scientists to choose what research they should pursue had become a very live issue with the advent of recombinant DNA and the potential of genetic engineering. It was held by some that research in these areas should be forbidden because of its possible effects on society; this view I could not, and still cannot, accept.

By 1979 the long-term effects of the sudden and prodigious expansion of higher education in the 1960s were becoming increasingly apparent and were being considerably increased by the financial cuts being imposed on universities in the United Kingdom by the University Grants Committee – cuts which were leading to a breakdown in the dual support system of university research which had in the past been one of the strong points of the British system of financing universities. These topics I dealt with at some length in 1979 (Appendix V). Although apparently unheeded at the time, it was gratifying to see that within a couple of years some movement in the directions I indicated in that Address occurred, and fortunately that movement is continuing. I am convinced that, especially in time of financial stringency, funds available for research should be concentrated on those centres and people doing first-class work and not dissipated over a large number of small units, many of them deficient in quality and contributing very little to progress. Such views are unpopular, but I believe they will in the end prevail; meanwhile, we are learning the hard way!

During my period of office as President I made a number of overseas visits in addition to the South American visit already mentioned. Some of these I made as an individual, e.g. to the Symposium on Natural Products in Moscow and Tashkent in 1978, and to a similar one on the Organic Chemistry of Phosphorus in Burzenin, Poland, in 1979, but others were on official business, involving, in many cases, the completion and signing of formal agreements between the Society and corresponding bodies in the countries concerned. In this way I visited the Soviet Union, China, the Philippines, Egypt, Japan and Jugoslavia, and made several visits to the National

Academy of Sciences in Washington – a body with which the Royal Society has always maintained very close relations. During the same period I also represented the Society at the Silver Jubilee of the Australian Academy of Science in 1979, and had the honour, in the same year, to receive the Lomonosov Medal of the Soviet Academy in Moscow.

Two visits paid to Iran during my period of office are worthy of special mention in the light of subsequent events in that country. The Empress Farah – a woman not only beautiful and charming, but very intelligent to boot – was anxious to see science properly organised and developed in Iran, believing that its disorganised state, and the lack of esteem in which it was held, were largely responsible for the emigration (mainly to the United States) of scientists whose loss Iran could ill afford. Her plans included the setting up of an Academy which would bring together the best scientists in the country, and be a kind of nerve-centre for development. To avoid any suspicion of corruption or intrigue in setting up such an academy and electing its first members, she turned to the Presidents of the Royal Society (myself and my immediate predecessor, Sir Alan Hodgkin), the National Academy of Sciences of the U.S.A. (Dr Philip Handler) and of the Japan Academy (Dr Kiyoo Wadati) and, in May 1976, we four were invited to visit Teheran and there, in concert with the Minister for Science and Education, we selected twenty individuals to be the founding members of the Imperial Iranian Academy of Science before returning to our respective countries. The Academy was then founded, and within a couple of years was so well established that Dr Handler and I were invited to attend its annual meeting in Teheran in October 1978. There we each delivered a Pahlavi Memorial Lecture (probably the first and last!) and were formally admitted as the only foreign members of the Academy by the Empress herself, who was its Patron. On my first visit, the only substantial city I visited outside the capital was Isfahan, but neither there nor in Teheran itself was there any sign of unrest; if there was unrest, it was sufficiently concealed to make it invisible to a visitor like myself, unfamiliar with the local language. On the second visit, things were quite different in Teheran; a considerable section of the work force seemed to be on strike, heavily armed troops were on duty, not just around the palace but also around public buildings and parks. There was almost daily trouble in the bazaar, and tanks and armoured troop carriers were a common sight on the main streets. Apart from inconvenience caused by the curfew, which was rigorously enforced, and by striking hotel staffs, the affairs of the Academy went quite smoothly as did our visit to the palace and to various receptions. I recall one of these latter given by the Minister for Foreign Affairs and attended by most of the diplomatic corps, at which I had a talk with a group of ambassadors about the situation in Teheran. They agreed that things were a bit difficult, but they gave it as their view that the army would deal with them and that everything would be back to normal in a week or two! Actually what happened in a week or two was full-scale revolution and expulsion of the Shah! On the day I left Teheran the roads leading out of the city were choked with traffic, and chaos reigned at the airport. I confess I was greatly relieved when I finally got on an aircraft and took off for Europe.

I do not know the present position of the Academy, having had no communication from it for some time. I do know that it survived the revolution for I had some correspondence with its administrative secretary, and indeed had a visit from its Honorary Secretary in, I think, 1980. I understood from him that although the word Imperial had been removed from its title the Academy had been allowed to carry on, and that I was still a member. How many Iranian members there are now I do not know; certainly quite a number of the twenty chosen originally by us went into exile. But they may well have been replaced by others.

I have already mentioned how, in 1973, I took over the chairmanship of the Nuffield Foundation. My predecessor resigned largely on the grounds of infirmity due to increasing age and, when I took over, I resolved that I should resign the chairmanship long before I was forced to do so by age and infirmity. Accordingly, when, in the autumn of 1979, I was approaching the age of seventy-two I decided to resign. This appeared to cause some alarm among my fellow Trustees, but I insisted, and formally announced my retirement as from 31 December 1979 (although I continued as a consultant and as an Ordinary Trustee of the Foundation). Much to the amusement of my family, this effort to divest myself of responsibility was not very successful, for, even before my formal date of retirement, I found myself involved in the creation and organisation of a charitable trust in Hong Kong of a similar size and with somewhat similar objects. It happened this way.

In the spring of 1970 my younger daughter, Hilary, who had set out to travel overland to the Far East with a view to taking a stage management job in connection with Expo 70 in Tokyo, ended up in Hong Kong where she took a job with a television company and remained there for about three years. Prior to this my wife and I had visited Hong Kong on several occasions, and we had good friends there in Sir Lindsay and Lady Ride. Sir Lindsay, a well-known figure in the Colony, had been Vice-Chancellor of the University of Hong Kong until his retirement in 1965. We also knew among others the professor of chemistry, Douglas Payne, who had worked in my department in Cambridge and who, with his wife Grace, was extremely kind and helpful to my daughter when she arrived in Hong Kong. Naturally, having a daughter there caused us to increase the number of our visits to Hong Kong, and we became frequent visitors from 1970 onwards, and continued to be such even after Hilary left and returned to England. Our growing association with Hong Kong depended not only on the Rides and numerous other friends we acquired there, but also, and perhaps more importantly, on my involvement with the recently created Chinese University of Hong Kong located near Shatin in the New Territories. This university, which occupies a magnificent site overlooking the sea, was formed as a bilingual university by an amalgamation of three colleges – Chung Chi College, United College and New Asia College – under the dynamic leadership of its first Vice-Chancellor (and effective creator) Dr C. M. Li, whom I came to know well. When I first saw the campus, the only college actually on it was Chung Chi, but now the whole university is located there and it is still growing. The Chinese University interested me greatly; as a bilingual institution which, unlike the older University of Hong Kong, could take entrants from the Chinese middle schools in Hong Kong, and operating four-year courses on a pattern closer to that used in the People's Republic of China, it appeared to have great potential for the future. With the passage of time the Republic and Hong Kong were bound to get closer and it seemed to me that the Chinese University could have a major rôle to play as a kind of bridge between the Chinese universities and those of the western world. Because of my belief in its potential, my contacts with it increased, and in 1978 I accepted an invitation to become a member of the University Council, a position which I still hold. In that capacity I have come to know and to enjoy the friendship of Dr Ma Lin who succeeded Dr Li as Vice-Chancellor, and who shares my view of the university's potential.

In 1978 I had a letter from Lady Ride (now, alas, widowed) telling me that she had an old friend, Mr Noël Croucher, who was in a very worried state and wanted my help. Noël Croucher had lived in Hong Kong since he arrived there as a very young man in 1910. A stockbroker, he amassed over the years a very large fortune; separated from his family, he lived alone in a large house part way up the Peak and, at the time of which I write, he had become something of a recluse. His fame in financial circles was legendary (I remember reading an article in *The Financial Times* in which he was described as the *éminence grise* of Hong Kong finance). Noël loathed publicity of any sort, and, although he had from time to time made substantial donations to schools and hospitals in Hong Kong, he took great pains to conceal as far as humanly possible the fact that he had done so. At the time Lady Ride wrote to me, he had been wondering how best to use for charitable purposes a substantial sum of money which he had, and did not wish to see unused. He had discussed the matter with some of his financial associates who had made various suggestions about setting up trusts, but he didn't really like any of the schemes proposed. Moreoever, he got an idea (probably quite erroneous) into his head that those to whom he had talked were really out to get hold of his money. When an old gentleman of eighty-eight gets ideas like that into his head they are very hard to remove; so it was that he talked about his troubles to Lady Ride, and decided he wished to discuss everything with me as someone he could trust. Within a very short time from my receipt of Lady Ride's original letter I had a long screed from Noël about his problems, and asking me to come and see him about them. Now, it is true that I had met Noël Croucher once or twice fleetingly at functions in Hong Kong, but until this time I really did not know him well and I was faintly alarmed when, in one of his letters, he added a P.S. to the effect that he didn't quite know how much money he wished to dispose of but the last time he had looked at it the sum was around £15 million (it transpired in the end to be much more than that). I was somewhat more alarmed when, after our first meeting, he seemed to have decided that I was the man for him and henceforth seemed unwilling to do anything about his proposed charitable activity unless I approved of it. He told me that his first priority would be to see that outstanding young Hong Kong Chinese graduates in science, technology, or medicine should be enabled to develop their talents further by post-graduate work in the United Kingdom or elsewhere in the British Commonwealth. That done, he would like to promote activities which would benefit Hong Kong and raise the standard of its higher educational institutions. For the rest, he was prepared to help promote cooperation with China but did not wish to be active in Singapore, Taiwan, or other strongholds of overseas Chinese where he felt there was already plenty of money if they would only use it properly. He did not want to have his money spent on bricks and mortar and he had no time for sociology and very little for social science in general!

Noël agreed that a Trust or Foundation should be set up and after some persuasion he agreed to be its first chairman, but only if I would be his vice-chairman and would undertake as such to get it going and to succeed him if anything should happen to him. He was readily agreeable to his lawyer, Ian MacCallum, also being a Trustee, and we persuaded him to add Dr Rayson Huang, Vice-Chancellor of the University of Hong

Kong and Sir John Butterfield, Master of Downing College and Professor of Physic in the University of Cambridge. Sir John was not merely eminent in British medicine but had, through the Hong Kong University and Polytechnic Grants Committee, substantial knowledge of medicine and science in Hong Kong institutions. In due course a Trust Deed setting up the Noël Croucher Foundation was drawn up and signed by Noël and his four Trustees in November 1979. A further Trustee in the person of Mr Michael Sandberg, chairman of the Hong Kong and Shanghai Banking Corporation, was appointed in 1980.

The reason for my family's amusement at this time is now obvious. In fact I had resigned the relatively simple task of being chairman of a well-established Foundation with an efficient administration only to find myself virtually in charge of a Foundation of not dissimilar size with no organisation whatever. To complete the story, I have to record that Noël, to our great regret, died suddenly following a massive heart attack in February 1980, leaving me with the task of getting the Foundation off the ground; the Foundation's size, and responsibilities were not diminished by the fact that, under the terms of Noël's will, it became residuary legatee of his estate. The history of the beginnings and the development of the Croucher Foundation will in due course be recorded; suffice to say now, that it has been in operation since Noël's death and has made a number of grants to the universities in Hong Kong and awarded a substantial number of scholarships and fellowships for research. I confess that, although it may have been tedious work from time to time, I have enjoyed my part in the Foundation's development and I believe we are operating in accordance with our Founder's wishes. Needless to say, being in charge of a charitable foundation with its seat in Hong Kong, I am now an even more frequent visitor to the colony!

As I approached the end of my five-year tenure of office as President of the Royal Society on 1 December 1980 I gave quite a lot of thought to the content of my final Anniversary Address. I decided to make it a vehicle for the expression of my views on the Society as it was when I became President and on the matters which, apart from my chemistry, had occupied me over many years, namely the relations between science and government in general, and the position in these matters which the Royal Society should occupy; I believe the best way to summarise here my views on these matters is to refer the reader to Appendix VI, in which I reproduce the relevant portion of my final Anniversary Address.

I thoroughly enjoyed my Presidency of the Royal Society, and it is my hope that I may have made some positive contribution in that position. It certainly provided me with an opportunity to influence, in some degree, action on many issues in public affairs in which I had been interested and involved over a period of some thirty years. The period between 1975 and 1980, however, is too close for me to give any considered judgement. When I look back further over my entire career, of which only a selective account has been given in these memoirs, it seems to me that I have been consistently fortunate in my family, my scientific work, and in my other interests. Chance, rather than design, has exercised a major influence – but that, I suppose, must be true for most of us.

APPENDIX I

'A Time to Think'. Presidential Address delivered to the Durham Meeting of the British Association on 2 September 1970. (Advancement of Science *1970,* **27,** *70*)

The impact of science on society has been discussed frequently and at great length by many people in recent years – so much so, indeed, that it would be difficult to put forward views on any single aspect of the subject which had not at one time or another been aired by someone else. Yet as a rule discussion has been fragmented; one speaker will talk about scientific manpower, another about science in developing countries, a third on how scientific advice should be made available to government, and so on. But the problems facing society today are so overwhelming that they demand our most urgent attention; in such circumstances the evidence we have regarding them warrants repeated discussion from all angles. In this Address I should like to take a rather broad view and to present certain thoughts and reflections which suggest, to me at least, actions which should be taken if we are to extricate ourselves from the social morass into which we, at times, appear to be sinking.

The first thing we must recognise is that science, although it has expanded enormously and with increasing speed during the past couple of centuries, has had of itself little or no *direct* effect on society. Nor could it have, since it is a cultural pursuit akin, indeed, to music and the arts; it seeks only to enlarge our understanding of the world in which we live and the universe of which our world forms a tiny part, using the experimental method which is its essential characteristic. What does affect society directly is technology. Not only does it do so now, but it always has done. From the moment when primitive man first used and fashioned tools and weapons, i.e. made his first technological innovations, his progress – and, indeed, his fantastically rapid rate of evolution – has been determined by technology, which is simply the application of discovery or invention to practical ends. The evolution of an animal species is, and must be, a very slow process when the species has effectively no control over its environment. This is where man differs from all other species – through technology he can consciously affect or control his environment, and down the ages he has been doing so to an increasing degree. There can surely be little doubt when one contrasts the development of, say, man and the chimpanzee from a common primate stock around five million years ago that the fantastic progress of one but not the other, especially as regards brain size, is hardly explicable save on the basis of a selection based at least partly on technological skills used in the control of the environment for man's benefit. I believe that man's evolution has been closely bound up with technological progress, and will no doubt continue to be so – for we must remember that, however much we may have interfered with the processes of natural evolution on the Earth, or however many species we have exterminated in the process, evolution is still going on and man as we know him today does not represent its end-point. At least, it will not unless man in his folly so misuses his mastery of technology as to destroy himself and the habitability of the Earth for any form of life as we know it. And the trouble is that man, by a series of enormous technological advances made in very recent times, has acquired almost unlimited power, at a time when his social progress gives no guarantee that this power will be wisely used. The nub of our problem today is fantastically rapid technological advance coupled with relatively slow social progress.

When we look back to the very beginnings of recorded history it is clear that our early forebears had already gone in for some potentially risky technological innovation. After all, to embark on changing one's environment by developing agriculture, or even by the domestication of animals, was really a pretty risky thing to do in the light of knowledge then available. No doubt man's early progress was punctuated by local failures and disasters, for there is an element of risk in every advance. Successes, however, clearly outweighed failures and as man's technology advanced so too did his social organisation, from the family to the village or tribe, and then on through city state to nation and federation of nations, each step forward being associated with new or improved technology in some sphere of human activity. Consider, for example, the way in which great empires have risen and fallen under the impact of improvements in military or administrative technology. During all this time – and I am speaking of the past five or six thousand years – each new technological advance brought in its train social changes, but the

pace of advance, although it gradually increased, remained on the
whole slow until the period extending over the late eighteenth and
early nineteenth century which is commonly described as the period
of the Industrial Revolution in the Western world. For this slowness,
compared with the fantastic speed with which we have seen technology advance since then, there were no doubt many reasons, but
I would single out the following three as being of prime importance:

(1) mechanical power was inadequate as long as man had to depend very
largely on muscle power, whether of himself or animals;
(2) communications were inadequate, so that the spread of any innovation
was very slow;
(3) technological advances depended entirely on the exploitation of
chance discovery or invention.

The pace of social change due to technological advances during
the historical period prior to the industrial revolution – although it
was irregular and subject to wide local variations – was, of course,
vastly greater than that of evolution which has, over millennia,
wrought enormous changes in all animal species. But it was, in
general, tolerable because it rarely involved revolutionary social or
occupational changes within the span of one lifetime. Man, like all
other animals, is essentially conservative; he seeks, above all, for
stability within his own lifetime. So it was that he slowly built up
a social system which was supported upon and legitimised by the
traditions of family, religion and of State, all of which combined with
a somewhat selective presentation of history to maintain a feeling
of continuity and security for the individual members of society.
Education rested largely on the tradition of apprenticeship; the child
learned from his father or other master a craft or trade, and on
completion of this training in early manhood he was provided with
a set of skills sufficient to carry him through the whole of his working
life. The age-old pattern of society – which, incidentally, entailed, for
many, hardships which we would now regard as intolerable – was
already showing ominous signs of cracking in the eighteenth century
in Europe, mainly because increased communication by sea had both
revealed new lands and brought Europeans more into contact with
other civilisations which had evolved along slightly different lines
from their own. But it was the Industrial Revolution which really
triggered the process of dissolution by undermining, in rather less
than a century and in succession, all three of the reasons I have
already mentioned for containing the rate of material change in our
society.

The invention of the steam engine was perhaps the most important feature of the early phase of the industrial revolution, for it put into the hands of man cheap mechanical power on a scale previously undreamed of. Not only did it revolutionise industry, but on land and sea it enormously increased the speed and scale of communications. New industries grew up, new materials of all types were imported into western Europe and from it a stream of machinery and manufactured goods flowed out to other parts of the world. Not only did the European Powers spread their influence over primitive areas such as America and much of Africa, but their aggressive new technology found the old and rather different civilizations of the Orient totally unprepared, and thus an easy prey to commercial and military aggression. The rapidity with which all these things occurred put an almost intolerable strain on existing societies, and unrest associated with the rise of an industrial proletariat was widespread by the middle of the nineteenth century. This, however, was but the beginning, for something new and vitally important to the development of technology happened round about the mid-century – something so important that I feel it could properly be said to have ushered in the Second Industrial Revolution, which has continued until today and whose far-reaching consequences we have not yet, perhaps, fully appreciated.

What we usually call the Industrial Revolution – I would prefer to call it the First Industrial Revolution – had little to do with science. It involved technological advances due to the exploitation of chance discovery or invention, and in this respect was no different from anything that had gone before. Men like Boulton and Watt were essentially inventor-entrepreneurs rather than scientists. Natural science had, it is true, been advancing steadily since the so-called scientific revolution of the seventeenth century, but although its theoretical basis and its corpus of knowledge were growing fast it was, prior to the mid-nineteenth century, mainly a pursuit of the amateur, and its impact on everyday life was small. But around the middle of the last century men of science began to apply the scientific method and the results of scientific research to the solution of industrial problems. Perhaps it is because I am a chemist that I am particularly struck by what happened in chemistry. In this country, for example, the young William Henry Perkin in the year 1856 produced a purple dye – mauveine – in the course of some over-ambitious attempts to synthesise the drug quinine. Perkin was not simply a scientist; he was also by nature an entrepreneur. Before this time the dyeing of fabrics had always been carried out with dyes like indigo, madder, etc., extracted from natural sources, but Perkin realised not only the commercial possibilities of mauveine, which he proceeded to exploit in a small factory at Hounslow, but he and others began deliberately to apply their chemical skills to the fashioning of more synthetic dyes of different colours. Thus were founded the great organic chemical industries of today, for it was from the dyestuff industry that the synthetic drug industry developed, as well as a host of others giving us entirely new materials – plastics, detergents, explosives, fibres and so on. This, indeed, was the birth of what we now know as technology – the application of science and the results of scientific research to the solution of practical problems: industrial, military, agricultural, medical and organisational. It is the new technology which has revolutionised our lives in this century and which has advanced at an ever-increasing speed, fed by, and itself feeding, a similarly advancing science.

I need not endeavour here to enumerate the sensational advances which have been made in almost every aspect of material existence, especially during the past fifty or sixty years, culminating in the staggering feat of man's landing on the Moon and his sending of further probes to other planets in the solar system. It is perhaps pertinent to remark here, however, that the part played by computer technology in space exploration, and the rapidly widening application of computers in industrial, administrative and medical work, may foreshadow yet a third industrial revolution with consequences as far-reaching as those of the second. The impact of these advances has been universal, but it has varied in scale and form in different parts of the world and has tended to increase rather than decrease the material gap between the rich developed countries and the poorer underdeveloped areas. In the growth of this gap the enormous rise in world population, which was an inevitable consequence of the Industrial Revolution and the development of medical science, has played, and continues to play, a major part. It is not now my purpose to discuss the problem of population control; suffice to say that it is a problem of the greatest urgency and one which man must solve if his aspirations to a better life in a world at peace are not to be stifled by sheer growth in his numbers.

Everywhere it is evident that economic strength, and with it national stability, are dependent on technological progress, and it is difficult to think of any area of national affairs in which science and technology are not involved directly or indirectly. Under such circumstances it is hard to believe that correct policy decisions can be reached in a democracy whose members are ignorant of science; and yet that is, even today, very much the position in this country. Educational patterns and social attitudes are closely related and both are slow to change for the reason I have already mentioned – the need which man feels to seek stability during his lifetime. So it is that while science and technology have been bringing about vast changes in our material existence at an ever-increasing pace we have failed to match them with appropriate social and educational changes. I believe the root of our present problems is in fact educational, that much of the frustration evident among young people is a symptom of our failure to adapt ourselves to rapid technological change and that we would do well to give educational change a high place among our priorities. Failure to do so could be disastrous, since the powers we now possess to alter drastically our environment are so great that, improperly or foolishly used, they could menace our own survival.

Advancing technology has brought with it great material benefits which have been widely spread throughout the entire population, so that the general standard of living has everywhere been raised and leisure increased. As a consequence there has been a gradual elimination of class barriers and a continuous movement in the direction of equality among members of our society. This has expressed itself in many ways, and not least in the demand for universal education which would give all children equal opportunity to develop their talents to the full. Response to this demand came first at the level of primary education, then moved on into secondary education, coupled with a rising school-leaving age, and now, with the added pressure of population growth, the demand is being made for universal tertiary education. There is really nothing surprising about the current push for tertiary education; it was quite predictable. But the general prolongation of the period of education to the age of twenty or twenty-one which it implies is, in my view, likely to present society with acute problems unless we can offer each individual the type of education appropriate to his or her future rôle in society. For the majority the education must have a vocational content such that they can proceed at the end of it to suitably lucrative employment without a further period of training akin to the old-fashioned apprenticeship.

There are two major difficulties in our way. The first is the failure to recognise that natural science is just as much a branch of culture as literature, music and the arts and that it is not in any sense a subject for the specialist only. In school curricula it should be part of the education given to every child and should be treated on all fours with the standard compulsory subjects like English, history and the rest. Unless this is done we will never have a scientifically conscious democracy. The second is that we must make available a diversity of types of tertiary education and must avoid over-emphasising (as we now do) the virtues of the traditional type of university education, especially when it is applied in a context for which it was not intended. Both of these problems are serious, but neither is insoluble. The necessary broadening of school education is gradually taking place, and the speed with which it is happening will increase as we get rid of current prejudices in the field of tertiary education. Over the last century or so we have built up in England a pyramidal system of education in which we started children in the primary schools at about five years of age and instituted from the age of twelve onwards a system of selection based on academic ability, under which those rejected at each successive selection step were channelled off into various technical forms of education or training. Each successfully completed step in this selection process was accompanied by increasing specialisation until, finally, at the top of the school pyramid there was left a small highly selected group which then went on to university. University education was, of course, highly specialised, but was designed, in each speciality, to offer to an intellectual élite the maximum opportunity to develop their intellectual and creative powers. Now, it is true that we did not operate the system very well, and that, in the past, social privilege may at times have been given more weight than academic ability; but at least that is the system which we built up, and I myself believe that, for the past forty years, during which secondary education has greatly expanded, it has been operating more and more on the basis of academic ability alone. During the past ten years or so we have seen a vast increase in the number of applications for university entry – so much so that we have roughly doubled the number of universities in order to meet the demand for places. Now we are told that we can expect a further doubling in numbers, so that by 1980 we are likely to have 450000 students in our universities. Such an increase, if it is to be absorbed, would almost certainly entail the creation of more new universities, although a good deal of the pressure could be relieved by increasing the size of the present universities, some of which are so small as to be barely viable, to around 10000 students each. But is this increase in the number and size of universities of the traditional type really desirable? I do not believe that the traditional type of English university education is appropriate for such a large proportion of each age-group; it was designed to deal with a small minority of our young people which was believed to be creative and to have powers of leadership. That such a group or élite exists must be clear to anyone who has been concerned with higher education, and it should, and indeed must, be given every opportunity to develop its power to the full. But that group is not going to amount to 450000 in 1980; what a large proportion of that 450000 ought to pursue is some other form of higher education with a different and greater vocational bias. Already the expansion of numbers in our universities has reached a point where the inappropriateness of the system to many of the entrants is evident. This is why we hear so much about bringing the universities closer to industry not only at the graduate level but also by giving a more industrial orientation to undergraduate courses in science, technology and (through management studies) the social sciences. We must remember that in this technological society we need a relatively small number of creative scientists and technologists to generate new ideas and a vastly greater number of technicians whose task it is to apply these to practical use. It is nonsense to suggest that both these types should have the same training, but I sometimes wonder if we pay sufficient heed to this problem. Somehow or other we seem increasingly to equate higher education with traditional university education and to regard the obtaining of a B.A. or B.Sc. as the goal to which all must strive. This 'degree fixation' has its roots in our own educational history but it has been powerfully reinforced by the argument that in the United States of America, one of the great 'super-Powers' of this century, the educational system is such that far more people take college (i.e. university) degrees than in this country. It is always difficult to make direct comparisons between countries with different cultural and social backgrounds, and there are some features of the American educational scene that are often overlooked in this country. One which I think is relevant to our problems is that although America has created a large number of university-type institutions giving Bachelor degrees of a rather general character and of a standard academically somewhat lower than the English Bachelor degree, with the increasing numbers flowing into higher education a marked hierarchy has developed within the system. Nowadays it is evident that a small number of universities are emerging as superior institutions working at a higher level than the others and with a large graduate element. It sometimes seems to me that those who advocate the adoption of a generalised American-type system of higher education do not realise that already America is moving away from that kind of system towards one which is more like the one we have built up slowly in this country. There is much talk of a new system of British universities in which there would be a two-year general degree followed by a further two-year specialised degree for a selected group and finally a very restricted two-year doctorate group. If something on these lines is adopted, as I believe it may well be, then we will have to come to some kind of hierarchical system in our universities – and perhaps it would be better to plan for a range of different types now rather than wait until change is forced upon us by circumstances. But whatever system is finally adopted I hope we will bear in mind that we need far more technicians than scientists and technologists. If we train too many of the latter then many of them will have to follow the career of technicians for which their training was not designed and which they will tend to regard as 'inferior' to that which they expected; the result will be a frustrated 'white-collar' class, with all the dangers to society that such a class implies.

I believe this frustrated class is, to some extent, already with us and that, together with the continued segregation of science in our schools and universities, it is in part responsible for the so-called 'drift from science' which is said to be visible in schools. This drift probably reflects also the 'anti-science and technology' attitude which in recent years has raised its head as a modern form of anti-intellectualism. This anti-science attitude originates, of course, in man's reluctance to accept radical change in his way of life; he is aware that technology is imposing such change, and if an individual has no understanding of science or technology he begins to regard science as some kind of monster over which he has no control whatever. People holding this view feel further justified by current concern over pollution and deterioration in the natural environment. There is no doubt that pollution of the environment by industry is a matter of grave concern to all of us. It is not, however, a necessary consequence of technological advance – it can be prevented. The fact is that pollution is not a new problem; only the scale is new. Hideous past examples of pollution and destruction of the environment abound – think of many of the industrial areas developed during the last century in north-east England, south Wales and in Scotland – but these tended to be ignored because they were local in their effect. Today the scale of industry accords not just with technological progress, but with the demands of an ever-increasing population, and so it is becoming evident that we must take serious steps to prevent or minimize pollution and undesirable change in our natural environment – steps which in the past we never bothered to consider. There are, of course, some difficulties about nature conservation; man always tries to alter the environment in his favour and it is inevitable that in doing so some of the changes he makes will be fatal to some other living species. This is something we must recognise, and we can only hope that in deciding whether interference with some part of nature is justified by the technological ends we seek, we will make our choices more wisely than we have sometimes made them in the past.

The making of correct choices, not only here but over the whole field of national and international affairs does, however, depend on the existence of an informed public opinion. In developing such an opinion, not only schools and universities, but also the British Association for the Advancement of Science, have their parts to play. It is interesting to see how the role of the Association has changed during its span of existence from the heyday of the First Industrial Revolution to the present day. In its beginnings, although the popularisation of science was borne in mind, its most important function was to provide a forum for the discussion of advances in science by those directly involved. With the passage of time this latter function has been largely taken over by the professional scientific societies; but as the Association's value in that sense has declined so its importance in spreading an awareness of science and technology among the public and particularly perhaps among its younger members has grown. That is why our BAYS activities seem to me to be so important. But since the young people of today are the adults of tomorrow we ought now to be giving serious thought to the pattern of the Association's other activities. Some changes we have made in recent years, but I doubt if they have gone far enough. It will not be a very easy task, for organisations, like individuals, tend to resist change or, at least, to insist that it be gradual. And there is the rub – technological advance and rapid change in our everyday life are synonymous. In the world of today and of tomorrow rapid change will continue, and if society is to survive it must so adapt itself that its members can not only live with change, but derive the maximum benefit from change. In talking around this subject I have emphasised the central position of education; but educational changes take time to show their effects and time is running short. Truly, for all of us, this is a time to think.

APPENDIX II

*Extract from Anniversary Address 30 November 1976.
Reprinted from* Proc. R. Soc. Lond. B. **196**, *7–12*
(1977)

Many of us can remember those difficult but heady days at the end
of the last war when it seemed as though science, the application of
which had achieved so much in war – radar, penicillin and the
atomic bomb to mention but three examples – would, in the brave
new world of amity and justice between nations symbolised in the
Charter of the United Nations, soon usher in the Millennium. Well,
as we are all only too well aware, the Millennium hasn't arrived yet
and, indeed, it seems in many ways rather further off now than it
did in 1945. Social and economic problems abound and many people,
especially the young, feel disappointed, frustrated and indeed let
down by the society in which they live. Now when people are frustrated they always look around for a scapegoat and I fear that far
too many cast science in that rôle today. This is not merely wrong;
it is indeed dangerous if it leads, as it has led, to a swing away from
science among young people entering higher education in many
advanced countries. For our standards of living cannot be maintained,
let alone improved, save through science or more precisely through
the application of the scientific method and the results of scientific
research to practical ends or, if you will, through technology and
technological innovation; this is as true of environmental as of
industrial problems. The real reason for most of our troubles lies not
in science but in our social and political ineptitude when it comes to
realising the potential of the advances which science has made and
continues to make. And so today we live in a turbulent and unhappy
world. The hoped-for spirit of amity among nations has failed to
materialise. Deep divisions exist between them and this has inevitably
led to increasing secrecy and mutual suspicion and all too frequently
to violence and even open warfare.

Secrecy has always been the enemy of scientific progress. This I hold to be true, but it is manifestly absurd in the imperfect world in which we live to appeal for the total abolition of secrecy and for the free and untrammelled circulation of all new knowledge. For example, one could hardly envisage the abandonment of all secrecy in defence research or free publication of all results obtained in the search for new drugs in the pharmaceutical industry. The existence of patents, of course, underlines the general acceptance of at least temporary rights of a proprietary nature in the results of research. In discussing secrecy I think it necessary to try to distinguish between two kinds of activity both of which are usually lumped together under the heading of research. The first, which is sometimes described as 'pure' research, is typically concerned with advances in our understanding of the natural world. As a general rule it is not undertaken in pursuit of any specific economic objective and it is characterised by a high creative content. From it come the new laws and hypotheses on which the progress of science depends. Since these require for their full establishment a consensus of opinion derived from widespread discussion and experimentation by many scientists it is clear that secrecy should be avoided at all costs in this type of research. The case is rather different with the second type of research, which we may call applied research and which often includes development – it is in fact the activity commonly described as research and development or, shortly, R. and D. Here the research is undertaken to solve particular problems, usually of an economic or military nature, with a view to technological innovation, and its most characteristic feature is ingenuity rather than creativity – ingenuity in the manipulation of existing knowledge and understanding. Of course, there is no clear-cut boundary between these types but broadly speaking the distinction can be made. In the world as it is, a measure of secrecy is usually inevitable in this second type of research; such secrecy will not actually promote the research, but it need not be unduly damaging provided freedom of exchange and discussion is preserved in the first type. As far as the individual scientist is concerned the type of research he undertakes should be a matter for individual decision, but having taken the decision he must abide by the rules.

All this may seem a statement of the obvious but I believe it needs saying at the present time. The danger of interference on political grounds with the free flow of scientific information and even dictation of the type of research which may be undertaken is still with us and may indeed be increasing. Persecution of a scientist because his findings conflicted with current religious dogma did not stop with Galileo, and the furore over Darwin's theory of evolution has not wholly died down even today. Yet it is fair to say that even two hundred years ago science had achieved a status which ensured for its practitioners quite a remarkable degree of tolerance and immunity from interference, always provided they followed the rules laid down for the Royal Society by Robert Hooke in 1662.*

The business and design of the Royal Society is – To improve the knowledge of naturall things, and all useful Arts, Manufactures, Mechanick practises, Engynes and Inventions by Experiments – (not meddling with Divinity, Metaphysics, Moralls, Politicks, Grammar, Rhetorick, or Logick).

So it was that two hundred years ago when America and Britain were at war, Benjamin Franklin, a Fellow of our Society and Founder of the American Philosophical Society, was able to obtain right of passage and freedom from molestation by American warships for ships of the Royal Navy under the command of Captain Cook engaged on a scientific expedition organised by the Royal Society. During the Napoleonic wars we know that Count Rumford travelled extensively in France, holding discussions with scientific colleagues, and it is also on record that in 1796 a French sailor and scientist, Chevalier de Rossel, at the time a prisoner of war in England, dined at the Royal Society Club as a guest of Alexander Dalrymple, the hydrographer to the Navy. As a final indication of the attitude of governments in the past one may quote the following excerpt from the instructions issued to the captain of H.M.S. *Rattlesnake* in 1846 (in which, incidentally, Huxley sailed as 'a surgeon who knew something about science'):

You are to refrain from any act of aggression towards a vessel or settlement of any nation with which we may be at war, as expeditions employed on behalf of discovery and science have always been considered as acting under a general safeguard.

I doubt very much whether such sentiments are widely shared by nations today and it is pertinent to ask why this should be so. I am no historian but I think the reason for the deterioration which has occurred is fairly clear. It had become recognised by the end of the seventeenth century that, particularly in matters concerning navigation, safety and progress were bound up with scientific discovery and invention; moreover, science was a common interest

---

* Cf. Weld, *History of the Royal Society* (London, 1848), vol. 1, p. 146.

of mankind and its discoveries were not associated or identifiable with any sectional interests in society. That it should have been granted a substantial measure of tolerance and immunity from interference by civilised communities is thus understandable. From about the middle of the nineteenth century, however, men began consciously to apply science, or more particularly the results of scientific research, to the solution of practical problems in agriculture, industry, medicine and defence. It was this – the new science-based technology – that enormously speeded up technological innovation and led to the fantastic and ever-increasing rate of advance in our material civilisation which has been the characteristic feature of the past hundred years. As a result science has come closer to its practical utilisation, and governments are increasingly interested in it. In its discoveries lie the seeds of power. The temptation to support and to control science in the interest of national political aims has therefore grown apace and the results are all too evident. Tolerance and freedom from persecution can no longer be taken for granted.

Flagrant examples of interference, like the promotion of the unsound ideas of Lysenko for political reasons with concomitant suppression of any work on genetics which might contradict them, may be rare, but persecution of scientists on political or racial grounds did not stop with Lavoisier. We have in recent times seen it happen in, for example, the McCarthy investigations in the United States, in the Soviet Union, in South America and before that in Nazi Germany. Closely associated is the danger that, even in the highly developed countries of the Western world, dictation of the nature of research permitted to individual scientists could develop not only from governments but also from militant, politically motivated minorities. Nor should we think that we in Britain are free from such intolerance. Not so long ago demonstrators in London prevented a well-known scientist from presenting and discussing his results simply because they believed that he might reach conclusions which would be at variance with their politically preconceived ideas.

Much of the creative work leading to fundamental advances in science is carried out in universities and research institutions. This work is for the most part uncommitted in the sense that it is not directed to any specific economic objective and, as it advances, its direction may and frequently does change as fresh areas of scientific ignorance are revealed. But today we are all too aware of the parrot-cry about the need for 'relevance' in academic research. Catch-phrases like 'cost-benefit analysis' and 'management of research' are bandied about and we are told of the need to orient research towards the fulfilment of national goals (these latter being, of course, determined by the particular political party which happens to be in power). This is ominously close to the direction of research on political grounds – a thing against which in an age of increasing political intolerance we must constantly be on our guard. The Royal Society, in accordance with the spirit of its Charter, has, throughout its history, sought to uphold the freedom of scientific enquiry and the exchange and discussion of experimental findings and theoretical ideas without regard to race, creed or national boundaries. But let us not forget that if scientists are to be accorded the privilege of tolerance and freedom from interference they must obey the rules. And these rules are most clearly expressed in my earlier quotation from Robert Hooke, in which he enjoins that there is no meddling 'with Divinity, Metaphysics, Moralls, Politicks, Grammar, Rhetorick or Logick'. To avoid meddling with some of these presents few problems but nowadays the scientist may find it less easy to keep clear of divinity, morals or politics. Yet if science is to lead to the advancement and not the destruction of mankind it must refuse to meddle with or be dominated by them. Science advances through the free interchange of experimental results and ideas and the main vehicles for advance are publication and open discussion. For the scientist therefore freedom to travel to attend scientific meetings and to confer or debate with colleagues at home and abroad is very important. Yet today arbitrary restrictions on such travel are not uncommon in some countries; moreover, they are often imposed at short notice and without any explanation being given. I would appeal to the governments concerned urgently to reconsider their attitude. For the welfare of their own countries will depend ultimately on the welfare of science; and there is no such thing as national science – no British science, no American science, no Soviet science – only science.

Refusal to allow a scientist to leave his own country and travel abroad is sometimes attributed to his possession of secret or classified information. This could be justifiable, in some cases, although surely only in a small minority. It is, however, extremely difficult to see any justification for the refusal of some countries to permit entry to scientists who have been invited to attend a scientific meeting simply on the ground that they are citizens of a country whose government pursues a political course which is unpopular with that of the host country. Perhaps even more deplorable is the way in which Unesco has decided to withhold financial support from any scientific meeting which allows participation by scientists whose governments are unpopular with a majority of its member states. Here is something which is in my view totally contrary to the spirit of the United Nations and is a threat to the freedom of science which should be resisted by every scientist, whatever his nationality.

During the past few years there has been much concern over the ill-treatment and indeed persecution of individual scientists in a number of countries by governments differing in their political stance. Many of us have been approached from time to time with the request that we sign a declaration or letter of protest about individual cases; such approaches have emanated both from private individuals and from organised bodies and the degree of documentation provided has varied. As President of the Royal Society I have myself – as no doubt has my predecessor in office – been approached with the request that the Society should associate itself publicly with such protests. But I have not acceded to these requests; for this reason I feel I ought to make my position in such matters clear.

Of the cases brought to my attention recently the commonest are those in which it is alleged that a Soviet scientist has been subjected to various forms of persecution including in some cases incarceration in a prison or psychiatric hospital for no reason other than that he, or some member of his family, has requested permission to emigrate. As presented, such cases represent the grossest violation of the Declaration of Human Rights embodied in the Charter of the United Nations to which all of its member countries have subscribed. The violation is, however, made neither better nor worse by the fact that the victim is a scientist rather than some other member of the community. Such infringement of human rights should be a matter for public condemnation and action by the United Nations. Alas, that organisation has not been very active in this respect; nor can one feel optimistic about it when one recalls the deplorable action taken by one of its organs, Unesco, over participation in international scientific meetings.

It is entirely right and proper that individuals should express their indignation about such cases by declarations or in any way they wish, but it is hard to see in what way the Royal Society can occupy a special position in the matter of human rights in general. I have already mentioned in these remarks the Society's three-centuries old insistence on freedom of scientific enquiry and discussion from direction or restriction on political, religious or racial grounds. This has again been made abundantly clear as recently as four years ago by its adherence to the Declaration on this subject issued by the International Council of Scientific Unions and by its cooperation with that body in efforts to uphold it in all its member countries. Moreover, in appropriate cases the Society has drawn and will continue to draw the attention of the Soviet Academy of Sciences or the corresponding body in any other country concerned, as well as our own Government, to the facts and to the need for action with, I believe, good effect. It is my firm belief that the Society as such can achieve much more in this way than it can by subscribing to or issuing public declarations. For it must be recognised that a scientist, as such, is in the same position as any other citizen of his country, subject to the same laws and having the same obligations to the society in which he lives. His profession does not entitle him to special privileges which are denied to his fellow citizens; nor should it deny to him those privileges which are the right of every man.

APPENDIX III

Extract from Anniversary Address 30 November 1977.
Reprinted from Proc. R. Soc. Lond. B. **200**, *x–xiv*
(1978)

A year ago in my first Anniversary Address to the Society I discussed
the problems of freedom in science both with respect to choice of
subject and the right of individual scientists to freedom of movement
and discussion with colleagues. Today I should like to examine in a
little more depth some specific problems which beset science and
scientific research in our country although science is not the only
facet of our society so beset nor are the problems confined to one
country. For this is above all a disappointing period in our country's
history. The coming on stream of oil from the North Sea basin does
not conceal the fact that the 'white-hot technological revolution' we
were promised never came to pass and the 'pound in our pocket'
is far from being what it used to be. In much of industry – but
fortunately not in all of it – there has been so little profit for so long
that some manufacturers are allowing their plants to run down and
are even skimping on the research and development which must
surely be essential to any regeneration of British industry. In the
country at large there is a widespread feeling that the legitimate
expectations of the early 1960s have somehow been frustrated. We
all know that we are less prosperous than we ought to have been and
it is cold comfort to observe that some other countries are equally
beset by inflation and ailing economies. This is a time of unsatisfied
aspirations of the young for opportunity and one too of growing
egalitarianism coupled too often with a lowering of educational
standards. Those of us who work in education and especially in
higher education cannot but sense the irony in the contrast between
government's enthusiastic launching of the Robbins expansion in
1963 and the stagnation of the present. Some there are who fear that
our universities may be irreparably damaged as a result of present

crises. I do not myself share that view for I believe our university system is at once too robust and too flexible to succumb. But what seems to me to be beyond dispute is that they, like many other institutions, will emerge transformed from the series of crises through which they are passing; whether for better or for worse remains to be seen.

In this, the Silver Jubilee Year of Her Majesty Queen Elizabeth II, it is appropriate that we should consider our successes and failures during the past twenty-five years and seek to profit from them in planning the way ahead. And our record in science and learning is an enviable one. We have indeed accomplished much in these twenty-five years and British intellectual life remains vigorous as ever; our scientific research is still flushed with success and full of promise. During this period no less than thirty Fellows of the Society have won Nobel Prizes and outstanding contributions have been made and continue to be made in many and diverse fields of science – in radioastronomy, astrophysics, chemistry, neurophysiology, plant genetics and molecular biology, to mention only a few. But in the present period of economic gloom the cry goes up – what has all this exciting work done for the country's economic problems? All kinds of people and, perhaps most ominously, many politicians ask why a country in such economic straits as ours should support academic research like this. Hasn't the time come when the universities should be harnessed to the regeneration of the British economy and the research in them devoted to the needs of manufacturing industry – to making better transistors or motor cars or even sewing machines?

Such criticisms are based on a profound misunderstanding of what universities are for and on a failure to appreciate what they have done for us since the last war. It is too easily forgotten that between the 1940s and the Robbins Report in 1963, government, industry and, indeed, the country at large was crying out for more and more trained scientists and engineers. In the chorus of complaint about their irrelevance to the country's economic needs it is too often overlooked that the universities have discharged their responsibility superbly in this respect. In less than a quarter of a century the numbers graduating in science and technology from British universities have multiplied by three and the numbers graduating with higher qualifications in technical subjects have increased still faster. And although many of these young men and women have been trained in institutions where people worry about such things as 'black holes' or bacterial genetics – for uncommitted research is a necessary adjunct to training – a very large proportion of them has found its way into the productive sectors of our economy. Indeed, for the first time in our industrial history, industry has enough technically trained people to satisfy its needs. Moreover we have now lived through the period when the burgeoning universities and poly-technics were two of the principal consumers of their own products – trained and talented young people. Unfortunately, as I shall discuss later, there is now a serious danger that the universities may be unable to recruit and retain the stream of young teachers and research workers on whom, in the longer term, their health and survival must depend. Those who claim that the universities have become 'irrelevant' forget that the universities have accomplished economically and without fuss the enormous task of expansion that they were set by the nation less than fifteen years ago.

Sometimes one hears complaints that there is something wrong about the education of scientists and technologists in our universities because many companies find that new graduates do not slip easily into their new rôles in industry. But it does not follow from this that there is anything radically wrong with our university courses; the acclimatisation of new recruits has always been a problem in industry and many devices have been tried to meet it. The reasons for the dissatisfaction that undoubtedly exists are at once more subtle and more complex. It remains true that the creative scientists and technologists of the future must be recruited from among those who have a broad grasp of what technology is about so that they can meet not just the present needs of industry but also those which industry has not yet foreseen. These are the people the universities seek to provide and it must be remembered that adherence to traditional ways and resistance to change for essentially social reasons are by no means uncommon in some sections of industry. It may well be that those graduates who are not going to proceed to a course of research before taking up employment would benefit from attendance at some kind of vocational M.Sc. course which would give them an introduction to particular features of industrial life. It could also be argued that in our rush to expand university education we have sucked into this stream of tertiary education a substantial number of those who might have been more appropriately trained through the more vocational education which it is the function of polytechnics to provide. But although minor changes should, and no doubt will, be made, I do not think there is anything fundamentally wrong with our university training in science and technology for those who are to play a leading rôle in industry and especially for those who are to provide the drive behind research and development.

This brings me back to the question of university research. This is a matter of major importance to the country's future, for research in universities has a double function. On the one hand, it has a training function which is vital to the development of creative scientists and technologists and which, although exercised primarily at the postgraduate level, also permeates the undergraduate years through its effect on the teaching and on the liveliness of the teaching staff. On the other hand, it has the function of advancing the frontiers of knowledge and the provision thereby of new facts and concepts upon which future technology must depend. If it is to fulfil this double function adequately academic research must be essentially uncommitted. This is not to say that it should have no objective – all research must be committed to that extent; but it should not be dominated by short-term practical or economic objectives. It is for this reason that proposals for joint industry/university Ph.D. courses or that universities should orient their research to meet specific industrial needs are, in my view, misguided. Industry is the proper place for industrial research; this, I hasten to add, is not to belittle in any way the contributions made incidentally by many university departments to industrial research and development. Similarly if government considers there is need for oriented research to support defence or meet the needs of nationalised industries it should make full use of its own research establishments for the purpose. There is often talk of a gap between certain industries and the university departments of science and technology related to them. But to the extent that it exists it is a gap of understanding which can be put right by closer personal contacts. It will not be closed by endeavouring to make university research departments do the job of industry or vice versa.

In Britain, unlike some other countries, most of our pure or uncommitted research is carried out within our universities or in units (as, for example, those of the Medical Research Council) closely associated with them physically and drawing on the same pool of young postgraduate students. And this is indeed one of the great sources of strength in our system. For in my view – and there is much evidence to support it – if research is to be of the highest quality over a long period it must meet one or other of two requirements. Either it must have definite (and preferably changing) economic objectives as, for example, in industry or in a mission-oriented government laboratory in which case it can operate with a mainly permanent staff or, if the objectives are not economic, high standards can only be maintained if there is a constant throughput of fresh young minds – a situation which obtains only in universities. I believe that it is largely because it has recognised this that Britain has maintained a consistently high standard and has such an outstanding record in scientific research; the system we have of concentrating most of our uncommitted research in universities is in fact one of our greatest assets.

That is why I and many of my colleagues are deeply concerned about the reduced level of support for university research which is apparent at the present time. It is, of course, clear that we must all tighten our belts and that there can be no question of continuing the fantastic growth in research funding which characterised the 1960s. The universities clearly must live with whatever level of research funding government is able to afford. But to do so we should be selective and concentrate on those centres where the greatest potential for progress lies. This is not an argument for or against so-called 'big science'; it is simply a recognition that when funds are not unlimited choices must be made and that these choices should not be made on the basis of 'big' or 'little' but on the quality and promise of the people in particular fields of science and on the likely pattern of demand for scientific manpower. At the present time university research faces two major problems. The first is the growing obsolescence of equipment in some sciences owing in part to spreading what money we have available for research too thinly over too many places, not all of which are of the standard we would wish them to be. The second, and perhaps in the longer run the more important, is this. As I mentioned earlier there was a period in the 1960s when the creation of new universities and the expansion of the old was proceeding so rapidly that they were consuming a large proportion of their own products to provide them with adequate staff. A very large number of permanent university posts were thus created and filled by young men and women of effectively the same age group all of them with many years of service ahead of them before retirement. On top of this the acute financial stringency experienced by the universities in recent years coupled with an easing of the pressure on entries has caused many universities to restrict severely the filling of posts rendered vacant by retirement. Such action, understandable enough from the university administrator's standpoint, has, of course, almost completely blocked the way ahead for many of the bright young men and women now coming forward who would normally have been absorbed into the university system at this stage in their careers and would thereby have helped to maintain its momentum. This state of affairs has in varying degree affected most faculties in our universities but its effects are especially dangerous in science. For these bright young scientists, although small in number, are the seed corn for the future of our industries. Without them, the research they promote, and the training they impart to their students, the necessary new knowledge will not be made available to regenerate and maintain British industry which will come to depend more and more on foreign know-how. And the danger is real: other countries see in our error their opportunity and in the absence of a real prospect of developing and exercising their talents here who can blame our young scientists if they take them elsewhere?

To identify a problem and to diagnose its origin is relatively easy but its solution or, better put, the amelioration of its effects is at once more difficult and more important. The major provider of money for university research is Government through the University Grants Committee and the Research Councils. These bodies are necessarily subject to a measure of political constraint although the experimental introduction of an advanced fellowship scheme by the Science Research Council shows that the existence of the problem has been recognised by at least one of them. The sums at the disposal of the Royal Society are very much smaller but even so I believe there is an opportunity for it to make a significant contribution. The Society has in its gift some eighteen research professorships and a substantial number of research fellowships of various types. We are hoping to use these professorships and fellowships as they become available to provide support for outstanding younger scientists and to build around them small groups of advanced workers so as to provide nuclei to initiate the scientific discoveries of the future and to retain and develop the best of our young people on whose work our technological future will ultimately depend. There are of course many problems to be faced and not least that of ensuring continuity. For the Royal Society cannot enter into a permanent commitment to maintain a group or unit. After a period of years – perhaps five or seven – a group supported in this way would either have fulfilled its aim and be ripe for dissolution or it should be taken over by and incorporated in the university where it is located. Those of us who have had experience of the takeover problem as between charitable foundations and the universities are well aware of past difficulties but I do not believe they are insurmountable. Only time will show if the efforts we are making will be successful; but I believe the attempt should be made and that it is in the best tradition of the Royal Society. In these days of rampant egalitarianism our concern for an élite in science may be regarded by some as outmoded. But it is not. In science the best is infinitely more important than the second best; that is the belief of the Society and a country which ignores or forgets it does so at its peril.

APPENDIX IV

Extract from Anniversary Address 30 November 1978.
Reprinted from Proc. R. Soc. Lond. *A.* **365**, *xii–xvii*
(1979)

Nowadays one often hears statements to the effect that civilisation is at a turning point and these statements are not infrequently coupled with a very gloomy outlook on the future of society or even with a denial that it has a future at all. Certainly it is true that there is much to discourage us the present scene. The subject was touched upon by the President of the National Academy of Sciences of the United States in his Presidential Report for 1977; he summarised the situation in the following words:

Consider the current scene: the largest, deadliest arms race in history, in a world that almost nourishes international tensions and conflict; self-defeating population growth in those nations least able to afford it; hunger and malnutrition on a vast scale; the countdown as domestic and foreign supplies of liquid and gaseous fossil fuels decline; uncertainty concerning the future of nuclear energy; pressure from the industrially less developed nations for a 'new economic order', generating ever harsher political strains; dependence of the industrial economy of the nation upon access to diverse mineral resources outside our boundaries, resources upon which we can no longer count simply because they are there; the economic consequences to this nation of the increasing industrial productivity of others; the new social problems attendant upon an aging population; the changing economic circumstances of various regions of our country; over-capacity of the nation's educational plant imposing constraints upon the career aspirations of young scholars; unsatisfied aspirations for opportunity, equity, and justice of various segments of our society; growing egalitarianism coupled, too frequently, with a lowering of educational standards; the twin spectres of unemployment and inflation; continuing decay of most of our cities; an inadequate but ever more expensive health care system; escalating costs of all services. Withal, we are sufficiently affluent to demand protection of the environment, both for aesthetic reasons and for protection of the public health, and to place ever greater emphasis on the safety of the materials, products and processes with which we traffic, introducing economic costs of considerable but uncertain magnitude.

These words were, of course, addressed to an American audience; but they could be addressed equally to a British one or to one from most other industrialised nations. And they certainly give food for thought. Change is inherent in progress (however one defines that term!) and so at all times people feel that there is something special about the particular period in which they live. Yet this is not necessarily so since our perspective of the present is distorting and the future is continuously being determined by us and by what we make of the present. Major transitions are rare although they do occur from time to time; one such was associated with the industrial revolution which began about two centuries ago and which has largely shaped the world we know today. That transition was, I believe, mainly due to one of the inventions that triggered the industrial revolution – that of the steam engine – which gave us access to plentiful and flexible mechanical power. All our tremendous scientific and technological achievements since then rest essentially on the stimulus given to society by that one invention. The social systems built up during previous centuries were unable to cope with the new circumstances of the industrial revolution and so there were many upheavals – some of them violent – from the French Revolution onwards during the period of flux before society, in the late nineteenth century, came to some kind of terms with the new world. But that accommodation could not last in the face of ever-accelerating technological advance and we are again in sore straits.

There are, it seems to me, so many similarities between the situation today and that of the early phases of the industrial revolution that, while acknowledging the difficulty of reaching an objective assessment of present events, I feel that we may indeed be living at another major period of transition. Again we have new inventions, all based, this time, on science, whose effects seem certain to be revolutionary and to impose severe strains – already becoming visible – on our society. Among the most vital of these new things are the harnessing of nuclear energy, the invention of the computer, the explosive development of micro-electronics, and the remarkable advances in molecular biology. All these have been proceeding so rapidly that during the past twenty years we have been brought face to face with a new world and are forced to look anew at ourselves and to adapt if we are to play any significant rôle in it. This is especially true of Britain which, although it was one of the first and most successful countries in seizing the opportunities presented by the earlier industrial revolution and in adapting its society to it, has not been outstandingly successful in this new one. There is probably no single or simple explanation for our economic decline relative to some other countries but I believe its origins are to be found in the latter part of the nineteenth century and lie in the twin effects of our early industrial success and the great development of the British Empire. I suspect that the vast inflow of wealth from the empire had a feather-bedding effect on our economy so that we were able to turn a blind eye to our growing industrial obsolescence and our declining productivity during the burgeoning era of science-based technology. And despite all changing circumstances we have gone on diminishing in our wealth-producing capacity and matters have been made worse by our failure to adjust our social and political systems to a rapidly changing world. It could well be argued, too, that a similar feather-bedding occurred in some other countries as a consequence of colonialism and the exploitation of the agricultural and mineral resources of the underdeveloped countries. Now that these under-developed countries – partly through population pressure – want a bigger share of the cake the shortcomings of more than one western economy are being revealed. The oil crisis of 1973 came as a rude shock to the industrialised countries and seemed at first likely to make them face their problems realistically. In Britain the discovery and exploitation of massive oil resources in the North Sea and adja-cent waters gave us a great opportunity – a kind of breathing space in which we could change our ways and build for a new future. I still hope we will seize this opportunity, although I sometimes fear that we may repeat our past disastrous behaviour and squander the proceeds of North Sea oil in propping up rather than reforming our antiquated economy so that before the end of this century we will be back in the mire again. There are disturbing signs that this may happen – deliberate overmanning and protection of jobs by subsidising lame duck industries rather than by the development of new industries and new jobs, low investment coupled with low profitability, and growth in public expenditure which seems to take little or no account of financial realities. In varying degree, of course, these or similar signs are visible in most industrialised nations and they have caused some people at least to argue that our civilisation is grinding to a halt and others to predict impending doom through exhaustion of the world's resources and inability to meet our energy needs. Personally, I cannot accept either of these gloomy predictions based as they are on what their proponents consider to be current trends. I have little faith in futurology, and forecasts of the future carried out by computer or crystal ball are about equally reliable. Of course the doomwatchers will be right if we do nothing and everything remains as it is now – but that is not, nor ever has been, the way the world goes.

The phenomenal rate of change which has characterised our material civilisation during this century has been wholly due to the application of scientific discoveries to practical problems – in a word, to science-based technology. Yet I wonder whether more than a very small fraction of the population ever pauses to think of the degree to which many of the accepted everyday features of our lives – automobiles, television, antibiotics and all the rest – have depended on science. Although none of us would want to be without these marvels – for that is what they are – some of us, it would seem, are so disheartened by all the social and economic problems we now face as to suggest that science is a hindrance rather than a help and that in the interest of mankind it should be controlled and regulated before it destroys us all. This is the view of the anti-science lobby which adduces the *Limits to Growth* thesis of the doomwatchers in its support and which vociferously supports extreme environmentalist views. The number of people dedicated to the promotion of such views is small but they obtain the support of a much wider section of the general public, including some of our politicians, who know little of science and who depend for their information about it on press, radio or television. In all these media the aim is to present information with maximum brevity and impact; inevitably this leads to the selection of sensational aspects of new discoveries which can be, and often are, dangerously misleading. Of course, no-one would claim that science has been a wholly unmixed blessing or deny that it has been on occasion misapplied. But on closer inspection its misuse usually turns out to be the fault of man and not of science – and often results from application by those too ignorant of science to realise the implications of its discoveries. At the same time one must admit that, sometimes, environmental problems like pollution have stemmed from short-sighted indifference to adverse effects on others which has all too often been manifest in the behaviour of governments as well as entrepreneurs.

I do not propose to argue here the rights and wrongs of (for example) pesticide usage or of the regulations surrounding the introduction and use of new products in medicine; much could be said about them but these are subjects for another occasion. What I wish to argue here is that just as we owe our present civilisation and standard of living largely to science it is only through the further promotion of science and technology that we will find solutions to many of the seemingly intractable problems set out at length by the *Limits to Growth* people. Thus I, for one, believe that the technical problems besetting the harnessing of thermonuclear fusion will be solved and mankind thereby given an inexhaustible supply of power. I believe too that the problems presented by diminishing natural resources could well be solved by the development of substitutes as yet unknown. This may sound a little like Micawberism, but it is not; of course we should take heed of the facts set out in *Limits to Growth* and be less wasteful of our resources – that is only commonsense. But if we continue to improve our natural knowledge all experience suggests that we will see changes which will radically alter the whole pattern of our lives – or if not of our lives then of those of our children and grandchildren; and we shall survive.

Since our future will be profoundly influenced by, if not wholly dependent upon, the degree to which we understand the world in which we live threats to the free development of science deserve close attention. I made some brief allusion in my first Address to the Society in 1976 to freedom of scientific research and the danger of political interference. Since then the situation has not improved and I make no apology for returning to the subject today. Ominously, voices have been raised claiming that limits should be set to scientific enquiry – that there are questions which should not be asked and research which should not be undertaken. These are matters which ought to be taken seriously the more so as they have not only been raised by members of the lay public but have even found support among some scientists. Currently the main focus of this attack upon the freedom of choice of the research scientist is to be found in biology. It is particularly marked in the area of molecular biology especially in relation to recombinant DNA, genetic engineering, the ageing process and the genetic component of differences in human beings.

It seems to me that the motives behind this questioning are of two types. The first is simple fear of disaster stemming from dangers inherent in the nature of the research or in the methods employed to carry it out. The second is more complex but is essentially ideological and includes quasi-religious objections; it sees in the new knowledge which is sought a threat to the established order of society or to the creation of a system predetermined in the light of some political dogma. In many cases both motives are mixed up with one another and it can be difficult at times to separate and identify them. A typical – and topical – example is to be found in the much publicised debates about recombinant DNA research. Since it involves the incorporation of genes or gene fragments from all kinds of organisms into a bacterium there to be transmitted indefinitely there is obviously a theoretical possibility of danger in such research. Those who call for its prohibition, claim that one might, in doing such work, accidentally create a new pathogenic organism resistant to all known antibiotics and might, again by accident, allow it to escape from the laboratory and cause a world-wide epidemic of some new and untreatable disease. (It is only fair to point out that as far back as 1974 scientists themselves pointed out the need to pursue recombinant DNA research under conditions of safety like those commonly employed in any research dealing with pathogenic organisms.) This is, like all such cases, one in which we have to balance risk with benefit, for no venture into unknown territory can possibly be without risk. Fortunately there is reason to believe that the common-sense view of taking safety precautions will prevail and draconian measures based on fears more appropriate to science-fiction will not be invoked. But it has been a stormy business largely because of confusion in the minds of many members of the public between recombinant DNA and genetic engineering. This confusion was very evident in the much publicised activities of the mayor of Cambridge, Massachusetts and his committee who sought to decide whether recombinant DNA research should be forbidden in their area and raised the spectre of the production of Frankenstein-like monsters through such work. Now, if indeed such monsters were ever to be produced, it would be done by genetic engineering which is not the same as recombinant DNA, although it is true that recombinant DNA research is an essential preliminary and will bring nearer the day when genetic engineering will be possible and could then be applied to deal with certain diseases. But why is it always the more horrific science-fiction aspects of as yet unmade discoveries that are publicised?

In questioning genetic engineering we are concerned not with safety but with ideology; applied to human beings it could alter the shape of things in a way which might not fit with preconceived ideas of the future. Objections to research on the ageing process are again ideological; if it were successful in greatly extending the life-span it could, the objectors argue, gravely upset the age-structure of the population and with it the whole nature of society. And studies on the importance of genetic differences in human beings are frowned upon because they might yield results which would conflict with political dogma. It is attempts such as these to control science on ideological grounds that are most dangerous and they must be resisted at all costs. Ideological control is complete negation of all that science stands for since it rests on the assumption that we know what the future will or should be or that we wish the future to be the same as the present; whether this is for socio-political or quasi-religious reasons is irrelevant. The fact is – as I have already stated – that we cannot predict the future of society on our present knowledge with or without computers, and no society can remain static and stable simultaneously. Science asks questions and on the answers to them our future depends. To forbid questioning is therefore unacceptable. There are also practical reasons why the control of science by regulating what it may and what it may not study is not even reasonable. Attempts to do so are almost certain to fail since the discoveries which lead to new advances in technology (which is what affects us directly) are made almost at random and frequently in areas of science which have no obvious relation to practical issues. I recognise that the *scale* on which scientific research may be pursued must be determined by economic considerations but I am wholly opposed to any attempts to regulate or control the *direction* of scientific enquiry and I believe that in saying so I also speak for the Royal Society. I also believe it to be important that the public should understand our point of view, and that we as scientists have been too reluctant to present our views publicly. Perhaps we should do more to correct false impressions and allay fears about scientific matters which derive from the methods of presentation currently employed in the public media of communication.

What I have just said refers to science; the situation is different when we consider technology. Technology is simply the application of discovery or invention to the solution of practical problems and it is technology and not science which has a direct effect on our daily lives. Today, of course, it is largely science-based but there is no reason why it should not be directed according to national interests. Moreover, some technological developments which could be undertaken on the basis of scientific discovery could well be undesirable and ought to be restrained. Not infrequently new and apparently desirable technology can pose questions which we are unable to answer because we lack scientific knowledge. What we do not know could well be more dangerous than what we know; that is particularly so in matters relating to our natural environment. Many of our pollution problems have their origin in past technological developments which were undertaken without knowledge of their potentially harmful consequences. Today, concern is expressed about possible effects of supersonic transport or the extended use of certain aerosols upon the upper atmosphere on a global scale – for we have advanced technologically to a point where our actions could have a global rather than a mere local effect. In this particular instance what we lack is scientific knowledge of the upper atmosphere and especially of its chemistry. Such knowledge should be sought and although, as I have argued, one cannot control the direction of scientific enquiry by decree it should be both possible and acceptable to encourage research on a topic of this type perhaps by increased funding. It is unfortunate, however, that much of the scientific work needed in the environmental field is not very exciting, requires an elaborate interdisciplinary approach and does not offer much scientific kudos to the individual investigator. How to get round these problems and attract into the environmental field a larger share of our best scientific talent is a major problem at the present time.

One of the most difficult problems governments frequently have to face is the choice between several alternative technological options; in some cases such as nuclear energy the choice could have widespread and important economic consequences. Choice is ultimately a matter of political and not scientific decision; but if the choice is to be wise it cannot be taken without scientific and technological advice. Here we approach the problems of science policy and the social responsibility of scientists. As to the latter the scientist has the same social responsibility as any other citizen; in discharging it, it is his duty to provide both government and the public with the facts of a scientific discovery or technological advance together with an objective appraisal of possible implications as far as he can foresee them. His task in a democracy is not to take political decisions, but to provide the evidence upon which rational decisions can be taken. That is why I believe that the recent activity of the Royal Society in promoting and publishing the findings of study groups and interdisciplinary discussions on current scientific problems and the issuing of reports and appraisals of Government reports on technological questions are so valuable; these activities should and I hope will be intensified in the national interest. But they ought to receive wider publicity and in this connection it may be that Fellows should be more ready than they, perhaps, have been, to make their views more widely known so as to combat misinformation of the public. For misinformation or slanted information is an everyday occurrence in matters scientific and it stems in large measure from the methods used for the dissemination of news. Abbreviation is the keynote and it reaches its peak in television where a snapshot-like visual and auditory effect is the objective; in striving for this, distortion in favour of the sensational or arresting is almost inevitable. I believe it to be very much in the public interest that an answer to this should be found.

APPENDIX V

Extract from Anniversary Address 30 November 1979.
Reprinted from Proc. R. Soc. Lond. A. **369**, *299–306*
(1980)

To me at least one of the most interesting features of the Report of Council is the evidence it provides of the Society's increasing concern with major problems and issues of the day where the provision of objective scientific evidence as a basis for political decision is necessary. Especially is it necessary in those matters where facts tend to be ignored or distorted by groups (often quite small) of ideologically motivated fanatics, or perhaps unintentionally by news reporters under the twin pressures of meeting a deadline and producing something which is at once brief and arresting. I had occasion last year to mention one such topic – recombinant DNA research. In this year's Report you will see that three new Royal Society Study Groups have been established: one on Assessment and Perception of Risks (Chairman, Sir Frederick Warner), a second on Safety in Research (Chairman, Sir Ewart Jones) and a third on The Nitrogen Cycle (Chairman, Professor W. D. P. Stewart). In addition a Joint Working Party on Biotechnology (i.e. the application of biological organisms, processes, and systems to industry) has been set up with the Advisory Council for Applied Research and Development (A.C.A.R.D.) and the Advisory Board for the Research Councils (A.B.R.C.) under the chairmanship of Dr A. Spinks. I would also draw your attention to the *ad hoc* group which under the chairmanship of the Physical Secretary is preparing a submission to the government's Commission on Energy and the Environment on the whole problem of coal and its future in the economy. The group is studying the available evidence on reserve identification and extraction, transport, and handling of coal as well as environmental effects, conversion and utilisation techniques and effluent problems; many of these important issues tend to be glossed over or ignored in public statements about a future (and hypothetical) 'coal economy'. Yet another *ad hoc* group under the chairmanship of Dr G. B. R. Feilden is considering afresh the interface between industry and the academic world and the rôle which the Society might play in the future development of the Industrial Research Associations. You will notice also the substantial number of Discussion Meetings that have been held. These meetings, which have been a feature of recent years, fulfil an important function. Not only are many of them interdisciplinary in their coverage but they can serve as a mechanism for focusing public attention on certain problems or matters of technical debate. It is my hope that our activities in these directions, including the preparation of reports on important national issues and responsible discussion of the problems involved – whether on our own initiative or at the request of government – will continue and expand. For, more than ever before, our daily existence is dependent on advances in science-based technology and our future depends more than many people seem to realise upon the use we make of the new technologies which will develop on the basis of today's discoveries in science. Failure to choose wisely among the various choices open to us, or, even worse, to ignore them in the vain hope of continuing to operate antiquated technologies successfully in the competitive arena of world trade spells disaster for any industrialised country. Yet this is what we have been doing in Britain in recent years although the extent of our economic decline is currently hidden from an unthinking public by the fortuitous (but temporary) inflow of wealth from the North Sea oilfields. Time was when the area of choice open to governments in the formulation of national policy was limited and the factors governing choice comparatively straightforward and simple to understand. But that time has long since gone. The development of science-based technology that followed on the heels of the industrial revolution continues to gather force and there is no way in which it can be halted. Human society cannot escape the consequences of new knowledge which will emerge from science in the future and, as the rate of accretion increases, so too will the complexity of choice and the number of options open to governments whatever their political colour. In a democracy like ours scientific expertise among politicians is hardly common and today governments are bombarded from all sides with a babel of advice from pressure groups, much of it misinformed or heavily biased. It is not my purpose today to argue in detail the mechanisms by which the need for external and independent advice should be met, but it seems to me that the Royal Society is a body uniquely constituted to organise the provision of that advice. I believe that this is an area in which the Society should be more active than it has been in recent times, and the setting up and further development of the study groups and discussions mentioned in the Report is an earnest of that belief.

Future developments in science and technology cannot be predicted; none of us can foresee the discoveries which will be made or the technologies to which they will give rise. All we can say – and that with some certainty – is that they will surprise us. But what we do know from the recent history of our own country is that the survival of a great nation and the standard of living enjoyed by its citizens depend on their ability and readiness to be in the forefront of new technologies as they emerge. And the best way of doing this is to be master of the science on which these technologies rest. In other words those countries will be the most successful which make discoveries in science and then exploit them through technology. Our record in discovery is good but during this century our performance in the highly competitive area of technological innovation has been, to say the least, disappointing. To recover the economic ground we have lost as a result will demand a greatly enhanced effort (and perhaps also a change of heart) on the part of our people. But any recovery – and we have the opportunity for one now through nature's gift of North Sea oil – will be of short duration if we cut back on our scientific research for financial or other reasons without considering the effect our economies may have on our future stock of scientists. There is, I fear, a good deal of evidence to suggest that we may even now be mortgaging our future in this respect.

During the past year the Officers and Council of the Society have been much involved in efforts to promote the development of research groups around promising research scientists. Such activity was foreshadowed in remarks I made in my Anniversary Address in 1977 and I am glad to say that we are now making progress. Although the amount of money available to us is small in relation to the overall need we hope the contribution we can make will not be wholly insignificant and may have the effect of encouraging other bodies to promote first-class research and raise the morale of those who are capable of doing it. For morale is still low in our universities and especially so among the younger members of the academic research community. The evidence for this is circumstantial and perhaps to some extent subjective but I believe the decline has now reached a point at which not only urgent consideration but also action is called for if not only our research but our whole university system is not to suffer permanent damage.

It is common to put the blame for this on the stagnation of the British economy and the consequent shortage of money for education and research. Our economic difficulties certainly play a large part but I believe the real root of the trouble lies in the misguided euphoria which in the early sixties caused us – in common with most other industrialised countries – approximately to double the number of our universities and greatly to expand our student numbers. No one would seriously dispute the thesis that higher education (tertiary education would perhaps be a better expression) should be available to all those able to benefit from it, but in those heady days higher education was equated with university education of the traditional pattern. How wrong this was has been amply demonstrated by subsequent events. The sudden expansion in student numbers involving as it did the entry of many with no real motivation for the type of education provided by our traditional universities (and which was, by and large, adopted by all the new ones) was in my view a material factor in the disturbances which marked the late sixties and early seventies. The universities survived the shock of these student disturbances surprisingly well but the long-term effects of the sudden and prodigious expansion of higher education are now becoming increasingly apparent. They are, in fact, basically demographic although in many respects exaggerated by our nation's economic decline.

The rapid growth of the university system in the 1960s brought about a vast expansion in tenured staff usually by the recruitment of relatively young men and women taken from the normal supply of university graduates and not invariably of the highest quality. Many, if not most, of these remain today at the same institutions and are likely to do so for perhaps another two decades under the tenure system which is almost universal. This phenomenon is, of course, not confined to the United Kingdom and it is giving increasing concern in most other industrialised countries, including the United States. The secretariat of the European Science Foundation has tried to collect and analyse such statistical material on university staffing as is available for a number of European countries (United Kingdom, West Germany, France, Denmark, Norway and Switzerland) and has published its findings in the Report of the Foundation for 1978. The results are strikingly similar for all the countries examined. Broadly speaking, they indicate that in most European countries the age distribution of teaching staff in universities is now at a level where 50%, or in some of them as much as 60% of the staff has an average age of about forty. The figures show, too, that in the countries examined in more detail (i.e. those listed above) the majority of those in post today will remain in office until the mid-1990s, when they will begin to reach retirement age. The demand for replacements during the next fifteen years or so will be exceedingly low and the effect could well be exaggerated if economic difficulties cause universities to economize by the short-sighted policy of suppressing posts that become vacant (this has, indeed, already occurred). The overall result both on research and teaching could be disastrous and persist for many years.

A detailed study of chemistry departments in British universities has been made by Professor Colin Eaborn who summarised his findings thus:

At present only 7.8% of the staff of chemistry departments are below 35 years of age; the proportion will probably fall to about 4% in five years' time and rise to only about 9% in ten years time. The proportion below 40 years of age, now 26% will fall to about 12.5% in five years and to about 10% in ten years time. During those ten years the proportion of staff over 50 will rise from the present 28% to about 62%. These figures have serious implications for British chemistry, and thus for the chemical and other science-based industries.

Indeed they have – and the situation in other physical sciences is unlikely to differ greatly. Viewing Europe as a whole, the European Science Foundation suggests that the chance for a junior research assistant to reach a permanent university appointment has gone down from about 70% during the 1960s to about 15% in this decade. The outlook is indeed bleak, faced as we are with a continued ageing of university staffs over nearly two decades and a denial to departments of the invigoration which younger recruits can bring to them. The other side of the coin is that younger scientists who would in previous decades have made a solid contribution to research in an academic environment will now be denied that opportunity and either abandon research or seek a career elsewhere. Surprising though it may seem, the academic research profession has within a few years been transformed from one of the most mobile to one of the most static. This is particularly so in the United Kingdom, where the economic stagnation of recent years has all but staunched the flow of academics in mid-career into other occupations. Many young scientists are seeking to retain their present precarious positions by hand-to-mouth grants because they fear that a change to another university or research institute might jeopardise their chance of obtaining a permanent appointment. This same attitude is probably responsible for the falling-off in applications received by the Society for fellowships under the European Exchange Scheme. To many, it would seem, the risk of losing a small chance which may exist at home seems too great for all but the most venturesome to follow the path of their predecessors, who normally received some part of their training abroad. In such circumstances it may well be that the diminishing chance of obtaining a university position could lead to a kind of negative selection process among those who stay at home, and in any case it militates against international scientific cooperation and understanding.

To these essentially demographic problems affecting our outlook in research others associated more directly with inflation and economic recession could be added. One only will I mention here; in the past, universities and research institutions spent something like 65% of their budget on staff and 35% on maintenance, including such things as research and library expenditures. As a result of inflation the proportion being spent on salaries and wages has been increasing and this has in many cases been reflected in a shortfall in research budgets, which has had to be met by increasing the demand on Research Council funds. How long such difficulties as these will persist it is impossible to say – they are likely to be with us until we emerge from the economic morass in which we are floundering. In these circumstances it is all the more necessary that the universities and those responsible for sponsoring the university system should grasp – and more courageously than they have been doing – the nettles which now abound, many but not all of them a consequence of underlying demographic problems. Several awkward, even painful, questions arise.

Is it, for example, to continue to be taken as an article of faith that all established academics are capable of first-class research and that all students obtaining first- or upper-second-class honours degrees should be encouraged and provided with the wherewithal to pursue academic research? The answer to both questions must, I fear, be no. Earlier in this address I indicated my view that the very rapid expansion of university staffs which has occurred was bound to involve some who were not of first-class research calibre. Moreover, we have seen in the past decade or so an approximate doubling of student numbers in our universities, the increase being drawn very largely from the same social classes of our population as in the past. Despite this the proportion of first-class honours degrees awarded each year in science at least has not diminished; it is difficult not to equate this with a lowering of standards. If these doubts are justified and we continue to believe that each department in every one of our present universities should have a substantial research school the strain on our financial resources may become intolerable. The resulting decline in the standard of our research will in due course extend through to our technology and so militate against the nation's economic recovery. Yet this is the road down which we appear to be travelling owing to over-rapid expansion and the inflexibility of our university employment patterns. The dual support system of the University Grants Committee and the Research Councils would ideally ensure that extra funds for the support of research were concentrated on those most likely to spend them well; as things now stand, however, while there is no reason to believe that really able young scientists with good projects to put forward will be denied temporary funding to initiate them, the prospects of their being given a sufficient measure of permanence to build up a centre of excellence in their institution is small indeed. Since in research only excellence begets excellence it is essential that universities individually or collectively should face up to this problem.

But can all university departments or even all universities really become centres of excellence? This question (and the inevitably disappointing answer to it) has been lurking in the minds of British academic scientists for the past decade but has not been openly faced. In reality, however, it is not possible – and it may not even be desirable – that all departments of, say, chemistry or physiology should stand out for the high quality of their research in some corner (and still less in all corners) of their discipline. Some, for example, have too small a staff adequately to sustain undergraduate teaching and to supervise at the same time programmes of postgraduate work which all departments in all disciplines appear to regard as the essential breath of life. Need this requirement for large postgraduate programmes be universal? It is a remarkable fact that in proportion to population there are more institutions awarding Ph.Ds in physics in Britain than in the United States. In the United States we find many more institutions devoted to professional education and applied sciences and there are many distinguished universities whose reputation rests substantially on the quality of their basic teaching.

Such diversity in our institutions should be encouraged. Why should there not be some differentiation between those who teach and those who pursue research? Even if by proceeding along these lines it should turn out that some of our universities become concerned essentially with undergraduate education of a more practical and vocational type than they now seek to provide, that would not be the end of the world.

Unfortunately the present machinery for the support of higher education and research in the United Kingdom was not devised for the encouragement of diversity. For all its undoubted virtues the University Grants Committee system does require that universities should compete within what is increasingly a common framework of objectives. These may be laudable enough as far as they go but with grants being geared to undergraduate numbers individual universities are not necessarily rewarded for doing well what they are best able to do.

The result is that competition between British universities is almost always competition on familiar terms – increasingly and unhappily competition at the margin for the increased tuition fees with which students are endowed by local authorities. Preliminary (and anecdotal) evidence suggests that in this competition Oxbridge and the older civic universities are winning out. If this be the case and the trend continues then we will end up with a hierarchy in which the universities lower in the pecking order will be trying to do the same things as the others but doing them with students who are not suited to them and would do better and become more useful citizens with a more vocational type of education such as is – or should be – provided by our polytechnics. Rather than let things drift slowly and painfully towards such an arid pattern why should we not seek boldly to alter patterns and develop diversity in our institutions now? At a time when a major effort is required to remain competitive in the technological revolution that is now occurring the idea that our traditional type of university education should be universally applied for reasons of social prestige is dangerous as well as foolish.

If we are to change our present university pattern the mechanisms by which we support research in academic institutions may also have to undergo some changes. Project grant applications will no longer be assessed simply on their merits alone without reference to the circumstances under which the research is to be carried out. Research projects in a given field will tend to be concentrated in one, or in only a very few, centres with considerably larger research groups than are usual today. It may well become necessary for the research councils to be more selective in the way in which postgraduate studentships are allocated to university departments; more radically, they may even have to think of making the grants to students of exceptional promise rather than to their potential supervisors. Another convention that may have to go is that every reasonably able Ph.D. graduate can expect as of right to have two or three years of postdoctoral research during which he can establish a claim on a tenured research post.

The frequent frustration of this expectation is one of the saddest of the current symptoms of malaise in our university research. Many postdoctoral fellows, many of them skilled and imaginative people, have discovered that there are more of them than of permanent jobs in what they have come to think of as their own field of research and must turn to something quite different. Nobody will deny that this entails a sad waste of skill; unhappily it is a circumstance that will not naturally go away until the British economy is once again buoyant. And the vigour of our research enterprise would surely suffer if all those concerned were now provided with a formal tenured career structure as many of them are now asking despite the current staff structure in our universities.

And so I come back to the over-rapid expansion of our universities in the 1960s and to the disastrous age-distribution in our university staffs which has resulted from the way in which it was carried out. The unfortunate circumstance that we have since then entered upon a period of acute economic recession has entailed severe restraint on the money available for teaching and research and led to resources being spread too thinly over too many centres. I have indicated in the course of my Address where I believe our problems lie and have only posed some of the questions which could be asked about research in our universities and the ways in which it is promoted. But in the last analysis it is difficult to see any real progress being made unless we can do something about our ageing university staffs and the lack of openings for our brightest young academic scientists.

Unfortunately there is no quick or easy answer. From their own resources universities themselves can probably do little and the breakdown of the quinquennial government-grant system makes forward planning well-nigh impossible for them. I believe that the efforts being made by the Society to support future leaders in research and the Science Research Council's scheme of advanced fellowships are valuable but in the aggregate they can provide only a small contribution to the solution of a very large problem. A reduction in the university retiring age to sixty without any reduction in pension would undoubtedly speed up return to a normal age-distribution, but whether in present circumstances government would be willing or indeed able to face the very large expenditure which would be necessary is doubtful. Encouragement of voluntary retirement at fifty-five with generous financial compensation has also been suggested but would probably not be welcomed by more than a few individuals, and even if it were generally acceptable it would be altogether too costly. Nevertheless I believe the problem must be tackled and that some of the unpalatable things I have said in this Address may help to point the way. For example, if we accept that there should be a kind of hierarchy in universities and that some of them will be much more devoted to vocational teaching and less to research than others, then only in a proportion of our universities need the situation be treated as urgent. These urgent cases could well have a retiring age of sixty (with full pension) introduced even if only temporarily so that a more normal flow of young academics could be reintroduced in them. This would certainly cost money but a great deal less than any blanket procedure applied to the academic system as a whole. I have not attempted a detailed calculation but I believe the overall cost would be tolerable. But it would involve the introduction of much more diversity into our university system than we now have, and this alone would make it worth while.

To do anything like this with our dual support system in its present form would be difficult and it is not surprising that under present circumstances many academics are beginning to ask whether the present system can continue. The more successful universities are probably right in thinking that they would secure a greater share of the resources available if they were able to compete within a more flexible framework and suggestions for change will certainly increase if the prediction of a decline in student numbers in the 1980s proves well founded. Perhaps it is not too soon to be thinking of the best form such a change should take.

APPENDIX VI

Extract from Anniversary Address 1 December 1980.
Reprinted from Proc. R. Soc. Lond. *A.* **211**, *6–13*
(1980)

In four previous Anniversary Addresses I have touched on a variety
of problems of current interest and importance which, although
matters of public concern, were in some of their facets of peculiar
moment to scientists. Today in delivering my fifth and final Address
to the Society as its President it is perhaps natural that I should look
back not simply on my period of office but also on the thirty-eight
years that have passed since I was elected to the Fellowship and reflect
on some of the changes which have occurred and on our situation
today. For changes have certainly taken place in the Society as in the
world outside it! At the time of my election in 1942 there were 460
Fellows and 48 Foreign Members; the number of Sectional
Committees was 8, of National Committees 9 and the total staff
numbered 15. Today we have 900 Fellows and 85 Foreign Members
with 12 Sectional Committees, 27 National Committees and our staff
numbers approximately 100. In the same period the number of
Fellows elected annually has risen from 20 to 40. This enormous
growth is of course a reflexion of the increasing fragmentation of
science and the large increase which has occurred in the number and
importance of scientists and technologists in this and other industrialised countries since the last war. With the recent increase
in annual admissions to 40 it is clear that for good or ill the size of
the Fellowship will be considerably larger than it now is before
anything like a steady state is reached. One obvious result of all this
has been that the Society has become more impersonal, and Fellows
living in areas remote from London have felt increasingly isolated
from its activities. In efforts to mitigate this Council has introduced
the *Royal Society News* and is now considering the possibility of

holding Discussion Meetings outside London. But other changes, some of them relating to the Society's concern with national policy, have occurred and it is perhaps instructive to look back at their origin.

When I was elected to the Fellowship in 1942 we were in the midst of a world war, and many of the activities in which the body of Fellows normally participated were either in abeyance or severely restricted. I had, as it happened, some basis for comparison because as a young research chemist in the thirties I had become much more aware of the Royal Society and its activities than most of my contemporaries through my father-in-law Sir Henry Dale. Sir Henry, who had been Biological Secretary from 1925 to 1935 and was to be President from 1940 to 1945, was, like many of his friends and colleagues on the biological side – men like Sherrington, Adrian, Hopkins, Mellanby, Barcroft and others – devoted to the ideals and traditions of the Society. To me in those days the Royal Society seemed like a rather exclusive gentlemen's club where occasional rather ill-attended meetings were held at which short scientific papers were read and after which the Fellows dined together at the Royal Society Club. In other words, it still had much of its original character after nearly three centuries of existence in London. In 1939 its main source of income was from private sources and the Parliamentary Grant-in-aid was £15 500. (For the current year the Grant-in-aid is £3.72 million and far outweighs our private income.) The Society had a few statutory involvements with government but these were not onerous and did not interfere with its essential independence. Even in those days it was recognised as the country's national academy of science, and as such acted as adhering body to the various international scientific unions which were in the early stages of their development in the decade or so before the last war. Its concern with public policy was limited until the exigencies of war thrust responsibility upon it.

The rôle which science should play in determining national policy has been the subject of almost continuous debate during the past thirty-five years and it is, in my view, relevant to any discussion of the position of the Royal Society today. The term 'science policy' which is widely used nowadays is, of course, a misnomer, but it is used umbrella-fashion to cover a variety of things which really fall under three headings – policy for science, scientifically based policy, and public policy determined in the light of available scientific information. Let me first try to exemplify them.

Science in its pure form, i.e. the improvement of natural knowledge as described in our Charter is, of course, a branch of culture just as much as music or the arts and to it as to these other branches government stands as a patron. In the case of science, however, it is not a wholly disinterested patron. For government is about power, and from science, or rather from scientific research, come discoveries in which lie the seeds of future power. Moreover, in a technological age the promotion of science is necessary in order that trained scientific manpower will be available to meet the country's needs. Government therefore is and must be prepared to devote substantial sums to the promotion of science. Of course, no government has unlimited resources at its disposal so that although it cannot – and must not attempt to – control the direction of scientific research it clearly must control the scale of expenditure and the weight of effort to be made in its various branches. A policy for science is therefore necessary. The second heading – that of scientifically based policy – is perhaps the one in which government involvement is of longest standing. It covers the promotion of activities involving scientific research which are essential to the national interest. In Britain the first example of this was the foundation of the Royal Observatory at Greenwich in 1675 by Charles II (although it was so grossly neglected by government in its early years that it would not have survived if the Royal Society had not taken it under its wing). The Observatory owed its creation to the manifest need for improvements in navigation which could only come through scientific research. Later examples are to be found in, for example, the Meteorological Office and the National Physical Laboratory. The third interface at which science and government come together, is where it is necessary to choose a policy or course of action from several alternatives among which choice involves not merely political and economic considerations but also a knowledge of scientific facts and their implications. Decision as to whether an energy policy should depend on nuclear power, on coal, on solar energy or on some other source of power is an example which is being widely discussed at the present time.

To understand the position of the Royal Society in such matters it is necessary to look back for a brief space at the changes which have occurred in the relationship between science and government during this century. The crucial factor in the enormous development of our material civilisation since about the middle of the nineteenth century has been science-based technology. Its growing importance naturally brought in its train an increasing demand for research and for trained scientific manpower. Universities and other institutions of tertiary education burgeoned and in them research, both pure and applied, grew in amount and became one of their standard features. Why the infusion of the new science-based technology into British industry should have lagged behind its introduction in some other countries during this period has been much discussed. I believe that an important factor in it was the feather-bedding effect of the enormous input of wealth from the Empire which concealed the growing obsolescence of our industry and our educational system and encouraged a false complacency. However, this is not the occasion to debate that topic interesting and important though it is. Whatever the reason, Britain was brought up with a shock on the outbreak of the First World War when it was found that she had become dependent on her enemy, Germany, for many of her needs – including, I have been told, even the dyestuff used for the khaki uniforms of her troops! Clearly action was called for and government set up a Department of Scientific and Industrial Research (D.S.I.R.) to promote science in industry at large. During the latter part of the war, too, in preparation for the post-war reconstruction of the economy, a committee on the machinery of government was set up under Lord Haldane – the Haldane Committee – and its recommendations set the pattern for government relations with science in Britain until the outbreak of the Second World War in 1939.

Briefly put, Haldane recognised that executive departments of government should have within them scientific organisations to ensure that research directly relevant to their needs would be carried out. However, because such organisations would inevitably be largely preoccupied with day to day requirements it would be necessary to have some other body or bodies which would be free from this and could promote scientific research of a longer term character. Initially these bodies were to be the D.S.I.R., the newly formed Medical Research Council (M.R.C.) and to them were added the Agricultural Research Council (A.R.C.) and much later the Nature Conservancy. Each of these was set up with its own laboratories and was charged also with the support of research in universities by means of student awards, creation of associated units and the support of researches 'of timeliness and promise'. These supplemented resources made available to universities through their general grant from the University Grants Committee and they really represent the origin of the dual support system for university research. In addition, D.S.I.R. was charged with the added duty of promoting research in British industry; one of the more interesting ways in which it sought to do this was by the creation of the Industrial Research Associations. In order to safeguard their independence and freedom from departmental influence or control D.S.I.R. and the Research Councils were placed under the Privy Council and their executive heads as well as members of their councils were appointed by the Lord President only after consultation with the President of the Royal Society. The position of the Royal Society as the country's national academy of science was recognised in this way but apart from occasional informal contacts between its President and Ministers it represented the sole involvement of the Society with the policies of government.

At the time of the Haldane Report and in the early years of the Research Councils it seemed that a fruitful relationship between science, industry and government was almost within sight. But that hope was not fulfilled; although matters were a great deal better than before, they still fell far short of expectations. Civil executive departments soon forgot about the desirability of having an active scientific organisation. Why, for example, should a Ministry of Transport bother about road research when the D.S.I.R. was there? If any awkward questions were ever asked it could use D.S.I.R. as a screen. Some of the more backward industries, far from being stimulated to do research, simply took the line that there was no need to spend much money on it since D.S.I.R. and the Research Associations would take care of it for them. Finally, the setting up of some government research establishmemnts under D.S.I.R. with permanent staffs but no challenging economic objectives to attain proved to be then, as it is today, a recipe for disaster. Despite such weaknesses, however, progress was indeed achieved during the inter-war period and even if the country was ill-prepared for war in 1939, its outbreak found Britain comparatively well supplied with operative scientific organisations which could be and did indeed become the basis for the enormous development of science as applied to the manifold problems of war between 1939 and 1945. The story of science in Britain during the last war is well known and need not be repeated here. Government, university and industrial research laboratories both jointly and separately made vital contributions – radar, penicillin, operational analysis and nuclear energy to name but a few. All aspects of science and public policy were involved and the central body which served as both the link with and adviser to government was the Scientific Advisory Committee to the War Cabinet. That committee consisted of the President and two Secretaries (A and B) of the Royal Society, and the executive Heads of the Research Councils under the chairmanship of the Lord President of the Council representing the government of the day.

When the war ended Britain was faced with tremendous problems; impoverished by its efforts, many of its cities devastated and its industrial economy distorted by the demands of total war the outlook was grim. But victory had been achieved, owing in no small measure to the sensational advances which had been made by science and technology and there was a feeling almost of euphoria – what science had done in war it could assuredly do in peace also. So the cry went up – let us have more scientists and technologists, let them have all the money they need and the millennium will be just around the corner. Given enthusiasm and some guidance from a scientific advisory committee like the one we had during the war, all would surely be well.

There is no doubt that at the end of the war the reputation of the Royal Society was high and its involvement with national policy greater than ever before; but these very facts faced it with a dilemma. What should be its future rôle? Three possible courses seemed open to it. First, it could have dropped all contact with government and reverted to being an isolated scientific élite with little or no influence on affairs – a pattern adopted by the national academies of the Latin countries and Japan. Secondly, it could have gone to the other extreme and become closely integrated as an organ of government with its officers holding political appointments; this is, of course, the pattern found in the Soviet Union, Eastern Europe and China. The third possibility was to adopt an intermediate stance in which the Society would retain its independence of government and avoid political involvement while maintaining informal contacts and being available to offer objective scientific advice as appropriate. It was entirely in keeping with Dale's passionate belief in the freedom and universality of science (a view reinforced by what had happened in Nazi Germany and the Soviet Union) that he chose the third of these possible modes of action; that choice was too, much closer to the tradition of the Society than any of the others. The resulting pattern has also been in varying degree adopted in Commonwealth countries, South Africa and Scandinavia. The National Academy of the United States although not integrated with government has much closer links with it than the Royal Society and carries out quite large-scale investigations or studies on its behalf.

Before it finally dissolved, the Scientific Advisory Committee to the

War Cabinet instigated the setting up of the so-called Barlow Committee to advise *inter alia* on the best way in which scientific advice could be made available to government at Cabinet level in time of peace. The Committee proposed that two bodies should be set up, an Advisory Council on Scientific Policy (A.C.S.P.) to deal with the whole field of civil science and technology and a Defence Policy Research Committee (D.P.R.C.) which for obvious reasons had to be a separate body. Under this scheme, which was in fact adopted, the link between these two bodies was provided by a common chairman, Sir Henry Tizard. As originally constituted in 1948 A.C.S.P. consisted of seven independent scientists and technologists from the academic and industrial worlds (one an Officer – not the President – of the Royal Society) together with an equal number of officials (secretaries of the Research Councils, chairman of the University Grants Committee and three others representing the Treasury, atomic energy and government science). When Sir Henry Tizard retired in 1952 I, who had been with Solly (now Lord) Zuckerman an original member of A.C.S.P., became its Chairman on a part-time basis with no personal commitment to the D.P.R.C. which had a separate chairman. (This position I held continuously until the dissolution of A.C.S.P. in 1964.) This seemed a very satisfactory arrangement at the time, giving as it did to the Royal Society a direct contact with the main civil science advisory body in government reporting to the Lord President of the Council who in those days was the Minister responsible 'for the formulation of government scientific policy' and indeed was given the added title of Minister for Science a few years later. The stage then seemed set for an effective system of advice to government in which the Society could play a role but which still ensured its essential independence and freedom of action.

Unfortunately the Society did not take full advantage of the situation. From 1950 under three successive Presidents the Society gradually lost influence and drifted away from matters of public policy; it became rather introspective and the Presidents were mainly concerned with such problems as accommodation, celebration of the Society's tercentenary and the like. This had unfortunate results in the early 1960s when a number of important – and in my view retrograde – steps were taken which radically altered the relationships between government, science and perhaps more especially the Royal Society. At that time there was much unease about the way in which Britain seemed to lag behind some other nations in technological innovation and there was a feeling that we were not making full use of the talent available in our rising generations because of inadequacies in our educational system. The Robbins Report recommending a huge (and to my mind ill-considered) expansion of higher education was accepted, *in toto* and almost without discussion, by both Government and Opposition in Parliament and the responsibility for science, the Research Councils and the University Grants Committee transferred to the new Secretary of State for Education and Science. The advent of a Labour Government in 1964 with its wild talk of a 'white hot technological revolution' completed the story. A.C.S.P. was abolished, technology was separated from science in a new ministry and a new Council for Scientific Policy was set up under the Department of Education and Science. Apart from recommending the pattern of division of available resources between the various Research Councils, this body had really very little function coexisting as it did with a Ministry of Technology with its own advisory council, and with the newly created post of Chief Scientific Adviser in the Cabinet Office. In these changes the Society took regrettably little part and its independence was in some measure affected by the political commitment of Lord Blackett to the Labour Government during his Presidency. Before my own election in 1975 several further changes occurred. The Council for Scientific Policy was dissolved and replaced by the more restricted but more more useful Advisory Board for the Research Councils, and following the resignation of Sir Alan Cottrell the office of Chief Scientific Adviser was abolished. Finally, changes – some of them not yet wholly absorbed – in the operations and interrelations of Research Councils and executive departments concerned with science, technology and medicine have occurred following the introduction of the so-called 'customer– contractor principle' adumbrated in the Rothschild Report of 1971.

When I assumed office I was unhappy about the fragmented state of science–government relations and the position of the Royal Society in that connection. The fact that the retirement of Lord Rothschild and the abolition of the Chief Scientific Adviser's post had left the Central Policy Review Staff without any scientific expertise at its disposal within government was a source of concern to its Chairman as it was to me, and in due course a scientific member was appointed to the C.P.R.S., much to its benefit. This appointment, although useful and indeed necessary, did not in my view provide more than an amelioration of our problems, most of which remained. I can, of course, give only a personal view of these problems and on possible ways of resolving the vexed questions of relations between science and government although I believe that view is substantially shared by my fellow Officers. To begin with, I hold that government needs a high-level independent scientific adviser who should be Chairman of an advisory council similar to the original Advisory Council on Scientific Policy. He could be whole-time or part-time but he should be independent of any department and should report direct to the Cabinet. Whether he should report direct to the Prime Minister is doubtful – Prime Ministers are likely to be so tied down by the day to day exigencies of government that it would probably be wiser to make science, technology and scientific policy the responsibility of a senior and influential Minister without Portfolio as it was in the days of the A.C.S.P. In the absence of an advisory body such as this which could call on the resources not just of departments but of the Royal Society and the Fellowship of Engineering, government will continue to depend on internal advisers from executive departments whose views must necessarily be in some measure partisan. What I here propose would entail the removal of responsibility for science and the Research Councils from the Department of Education and Science; I believe such a change would be in the best interest of science which must inevitably play second fiddle to education under present arrangements. It would moreover make easier and more effective the revision of our dual support system for research in universities which is sorely in need of reform.

These being my views it is only fair that I should indicate whether I and the Society have been able in any way to assist their promotion and, if so, to what extent. Following the appointment of a scientific officer to the Central Policy Review Staff I participated in a number of discussions and arising in part from these government set up a new body called the Advisory Council for Applied Research and Development (A.C.A.R.D.), a body consisting largely of independent scientists and technologists from industry and the universities with the Lord Privy Seal as titular chairman and a Fellow of the Society as operative deputy chairman. The formation of A.C.A.R.D. represents a considerable step forward; not only does its membership include several Fellows but the Society has collaborated with it and with A.B.R.C. to produce a most valuable report on Biotechnology, some of whose recommendations are now being put into effect as part of national policy. A.C.A.R.D. has also produced several other smaller reports and its actions to date would seem to augur well for its future. True I consider that A.C.A.R.D.'s remit should extend further than 'applied research and development' if it is to achieve all I would hope for, but until it is accepted by government that there should be a separation of the Research Councils (or, if you will – science) from education some limitations on its activities in regard to scientific policy will remain. In parallel too with the activities of A.C.A.R.D. the Society itself has undertaken, in some cases at the request of government, impartial reviews of evidence obtainable on, for example, the outlook for a 'coal economy'. In this and other ways it is maintaining and again increasing its informal contacts with government in the area of scientific policy.

Progress in such matters is of necessity slow but I feel we are moving on the right lines and that in doing so we not only uphold but maintain for the future the position in our country's affairs that was sought, and in substantial measure achieved, by our predecessors of thirty-five years ago. The Society's objects are and must remain threefold:

(1) To protect and encourage science in all its aspects pure or applied. As Robert Hooke once put it: 'To improve the knowledge of naturall things and all useful Arts, Manufactures, Mechanick practises, Engynes and Inventions by Experiment'.

(2) To offer to government an independent source of advice and help in the creation and operation of instruments through which science and technology may be brought fully to bear upon the formulation of national policy.

(3) To uphold and develop international scientific relations upholding the principle that scientists shall be free to interchange their findings and to collaborate in the search for knowledge without let or hindrance.

To realise these objects the Society must continue to maintain its independence, avoid involvement in politics and at all costs maintain its high standards. The Royal Society is and must remain an élite body if it is to retain its prestige and even its credibility.